DISCARD

How the Earthquake Bird
Got Its Name
and Other Tales of an
Unbalanced Nature

H. H. Shugart

How the Earthquake Bird Got Its Name and Other Tales of an Unbalanced Nature

Yale University Press
New Haven & London

Set in Nofret and Simoncini Garamond types by The Composing Room
of Michigan, Inc.
Printed in the United States of America.

Library of Congress Cataloging-in-Publication Data

Shugart, H. H.
     How the earthquake bird got its name and other tales of an unbalanced nature /
  H. H. Shugart.
        p. cm.
     Includes bibliographical references and index.
     ISBN 0-300-10457-X (alk. paper)

     1. Animals—Anecdotes.   2. Ecology.   3. Global environmental change.
  I. Title.

  QL791.S61152 2004
  590—dc22

                                                                        2004051648

A catalogue record for this book is available from the British Library.

The paper in this book meets the guidelines for permanence and durability
of the Committee on Production Guidelines for Book Longevity of the Council
on Library Resources.

10   9   8   7   6   5   4   3   2   1

For Brian, Lyndele, and Ramona
in appreciation of their support

Contents

Preface

Etosha, Tongaroa, Kakadu, Okavango, Denali—"The Last Wild Places" all seem to have beautiful names from another time to match their own wondrous beauty. Surely "The Last . . ." is a melancholy beginning. It is an epitaph more than a title. Whether in The Last of the Mohicans, The Last of the Summer Wine, The Last Picture Show, or The Last of the Wild Places, the phrase signals bitter reconciliation to a sad but inevitable end. I have visited and even had the good fortune to conduct research in some of the locales that tourist brochures proclaim as "The Last Wild Places," enticing us to see them before they are gone. I wrote this book because I do not wish to accept their leaving as inevitable.

Human activities are altering the planet—its atmosphere, its oceans, and, as I emphasize in this book, its landscapes. It should not come as a great surprise that we are living in a time of Last Wild Places and attendant high extinction rates of plant and animal species. Directly or indirectly, we are the cause. Understanding how to manage and preserve some of these places and their species can be remarkably difficult. But the causes of the high rates of extinction are easy to understand. Fundamentally, they involve change and responses to change, and these are the focus of this book.

To help you understand the impact of change on terrestrial landscapes, I will introduce ecological concepts describing the interactions between plants and animals, the relationship between landscape patterns and environmental change, and the connections involving extinctions and explosions of species populations. Each chapter begins with a quotation from mythology or history and tells the story of a particular animal and its fate. That story then leads to discussion of an ecological concept.

Why use animal stories? People in diverse cultures are educated in the ways of life through stories—tales to learn to speak, fables to discern proper behavior, myths and parables to remember truths. Often these narratives involve animals. Some animal stories give these creatures attributes exemplifying human traits to educate, as in Aesop's fables, or to entertain, as in Rudyard Kipling's just-so stories. My own stories are nonromanticized, historical accounts of certain specific animals.

One could undoubtedly write a comparable book using aquatic and marine animals to illustrate the effects of human change on our planet. I have restricted myself to terrestrial animals, either birds or mammals—five of each. Although they have the advantage of being familiar topics of traditional fables, they are an extremely thin slice of the rich diversity of terrestrial animals. My selections may seem diverse and eclectic, but each makes an important point about how landscapes function and how change affects the animals inhabiting them. I consider mostly North American species, but representatives from Europe, Africa, Antarctica, Australia, and Oceania are also present.

Perhaps one advantage of using animal stories is that the chapters parallel the intellectual evolution of many scientists performing research in large-scale ecology. Certainly my own keen interest in birds as a young boy led rather naturally to a desire to understand the factors that change the habitats of birds, and from there to an interest in ecosystems and global ecology. In talking with colleagues, I have often found that an interest in natural history, followed by realization of the fragility of species survival in a changing world, is one life pathway that produces global ecologists. It also can fuel an ardent involvement in conservation issues.

Some of my accounts are about familiar beings: dogs, penguins, and beavers. Other creatures are perhaps less familiar: extinct birds from the now-cleared floodplain forest of the Mississippi River, giant flightless birds, and an extremely common bird that most of us do not know at all. If there is a personal signature in the choices, it is my use of the Bachman's warbler ("the Earthquake Bird") and the ivory-billed woodpecker. These are birds that I long sought as a boy, but today am sure

I will never see. They were the rarest of the remarkable creatures of the forests of the Mississippi River and other meandering rivers of the southern United States. These vast forests had been wounded by the cotton farming of my grandfather and his peers, chained as my father's generation tamed the floodwaters that drove them, and destroyed by the value of soybeans to my own contemporaries. These wonderful forests are gone in the span of but one generation of the trees that made them. Their story will be repeated from the Amazon, the Zambezi, the Bhramaputra—unless we think deeply about what we are doing to our planet and steer it by a truer star than immediacy.

Although the implications of this book with respect to ecosystem change and species extinctions may seem grim and pessimistic, the experience of writing it has left me optimistic that a better appreciation of the scientific issues by thoughtful and caring people is our best hope for the future. We must be aware of what we do know, so that we can better define what we need to know. It is my hope that this collection of stories will provide some of that baseline knowledge.

I have been given considerable help in developing this project. Much of the initial writing was done during a sabbatical visit from the University of Virginia to the Research School of Biological Sciences at the Australian National University (ANU) and the Commonwealth Scientific and Industrial Research Organisation (CSIRO), Division of Wildlife and Ecology (now the Division of Sustainable Ecosystems), both in Canberra, Australia. The librarians of both organizations allowed me repeatedly to drain their shelves of books. They also helped me locate some of the harder-to-find material. Later, the libraries at the University of Virginia, particularly staff members Melissa Loggans and Linda Cotton, provided support and encouragement.

I appreciate the backing of the U.S. National Aeronautics and Space Agency (NASA grants NAG-7956 and NAG-9357) and of the University of Virginia's internal grant programs (the Academic Enhancement Program and the Funds for Excellence in Science and Technology Program).

Writing a book causes one to depend greatly on the help of friends and the kindness of strangers. My sincere thanks go to all those who assisted me with the writing. A. E. Newsome of CSIRO read my pieces about dingoes and discussed his experience with them over thirty years of research in the Australian outback; C. K. Williams provided similar aid on the European rabbit sections, and on the history and future of biological control of rabbits; Steve Cork discussed the endangered Leadbeater's possum and steered me to David B. Lindenmayer, whose work with this highly endangered Australian marsupial is the epitome of quality research on rare-species conservation; Monika Van Wensveen gave me access to her scanning equipment and help with obtaining photographs of Australian animals.

R. W. McDiarmid and A. P. Peterson helped me solve the puzzle of exactly which auk, strangely called *Alca pica,* was likened to the penguin by Joseph Banks. This research ultimately involved knowing which edition of Linnaeus would have been in circulation when Banks departed England on James Cook's HMS *Endeavour* voyage of discovery. R. D. Barber provided much-appreciated advice on the Bachman's warbler.

D. L. Druckenbrod, J. A. Blackburn, J. F. Lamoreux, R. K. Shugart, several reviewers on the Yale University Press advisory board, H. E. Epstein, and the twenty-one fourth-year students in the spring 2001 seminar in terrestrial ecology read the manuscript in its entirety, providing advice and encouragement along the way. Alan Newsome, Kent Williams, Erika C. Shugart, R. J. Swap, Mike Erwin, and Todd Dennis all gave their time and effort. So did Deborah Lawrence's 2003 seminar in conservation ecology. I am indebted to them all.

My gratitude goes also to Jean Thomson Black and Jenya Weinreb of the Yale University Press, who worked patiently with me during manuscript development. The assistance of R. L. Smith, Jr., in preparing the figures is greatly appreciated, as is the editorial help of Vivian Wheeler.

Three people helped me immeasurably, and in different ways. Brian Walker, an ecologist whose advice I have always valued, encouraged me for years to write a book about landscape ecology. The project would have probably never seen completion without the loyal efforts of my colleague Lyndele von Schill. Finally, I extend warm affection and appreciation to Ramona, who in our thirty-seven years of marriage has listened to my ideas, traveled the world with me, and tolerated the vagaries of living with a research scientist trying to write a book.

1 Introduction

> What, is the jay more precious than the lark,
> Because his feathers are more beautiful?
> Or is the adder better than the eel,
> Because his painted skin contents the eye?
> —Petruchio in act 4, scene 3 of "*The Taming of the Shrew*"
> William Shakespeare (written circa 1594)

Petruchio's dilemma is our dilemma as well.[1] We choose whether the adder is better than the eel: in the decisions we make that alter the planet's environment, we favor some species and disfavor others. Intentionally or unintentionally, we manage the biota and the environment of our planet. Our decisions have eliminated some species and created the opportunities and environments for others to become pests.

However, our challenge is more complex than deciding which species to favor in our planetary management. In a changing and interactive world, we are unable to do only a single thing; the interactions in ecosystems cause one change to produce another change and yet another. The question is how far these may carry forward in a progression of additional changes.

In times past, the seeming connectedness of natural systems made scientifically inclined intellectuals see the universe as a well-assembled clock. The wonderful interaction between form and functioning that was obvious in the natural history of plants and animals eventually implied that extinction of species was not included in the workings of nature. The world seemed part of an intricate clockwork universe. Nature appeared to work very well; indeed, it functioned as a clock built by a divine clockmaker. A clock with missing parts does not work; natural systems do work. Therefore, natural systems must not have any missing pieces and extinction must be unnatural.

This view is embedded in Thomas Jefferson's description of the fossil bones of a gigantic ground sloth that he found and named the megalonyx.[2] Because of its large claws, he took the fossils to be those of some sort of oversized carnivore (*megalonyx* ≈ large + lion). Jefferson felt that the megalonyx was still alive somewhere in North America and reasoned: "The bones exist: therefore the animal has existed. The movements of nature are in a never-ending cycle. The animal species which has once been put into a train of motion, is probably still moving in that train. For if one link in nature's chain be lost, another and another might be lost, till this whole system of things should evanish by piecemeal . . . If this animal has once existed, it is probable that . . . he still exists."[3] When Lewis and Clark were sent to explore the Louisiana Purchase in 1803, President Jefferson instructed them to watch for and report on large beasts such as the megalonyx, which he thought likely were alive somewhere in the American West. Unfortunately, the megalonyx and a remarkable diversity of other large mammals that were its contemporaries were already extinct.

Jefferson's clockwork universe theme is echoed today by familiar phrases such as "the balance of nature," "the wilderness concept," or "the virgin forest" and "the unspoiled prairie." We manage parks, nature reserves and conservation preserves with these concepts as the foundation of our actions. The objective of this book is to provide an alternative view, to give insights into the dynamically changing nature of ecosystems and the implications of this dynamism for our stewardship of the planet.

Extinction of species is a part of the Earth's biological history. Since higher forms of life evolved, periodic catastrophes have been associated with mass extinctions. Nine such extinctions may have occurred in two combined cycles. While the topic is hotly debated, these cycles conform to large changes on the Earth's surface (volcanic eruptions and other tectonic events, sea-level changes, reversals in the magnetic poles) and evidence of prehistoric meteor impacts. These cataclysmic events may derive their periodic nature from the movement of our solar system with respect to the plane of the Milky Way galaxy.[4] One of these cycles has extinctions occurring every 33 million years or so; it is combined with a longer cycle occurring every 260 million years.

The largest extinction "event" occurred 245 million years ago (in the Permian period), when marine animal families dropped by more than 50 percent and the number of known genera was reduced by around 80 percent. Such drops in the existence of genera imply even higher losses of species, perhaps as much as 96 percent.[5] Another extinction, better known to many, occurred at the end of the Cretaceous period, 65 million years ago. Often called the "End of the Age of Dinosaurs," it involved the extinction of approximately half of the living genera: many kinds of microscopic aquatic plants and animals, marine and flying reptiles, and dinosaurs.[6] However, land plants, crocodiles, snakes, mammals and many kinds of invertebrates survived these mass extinctions.[7] There are (and likely will continue to be) differences of opinion as to the cause; an asteroid impact or violent volcanic eruptions have been suggested as possible suspects.[8]

We know that extinctions do occur, sometimes on an incredible scale. We are still trying to understand whether or not ecosystems behave more or less similarly when the diversity of species is altered. With enough extinctions, does the whole system, "evanish by piecemeal"? The relation between the species richness of ecological systems and the large-scale performance of these ecosystems is difficult to study. Given how little we know, we have produced a remarkable volume of speculation about the connections.

One might argue that in such a dynamic world any extinction that could substantially alter ecosystems should have already done so. In a world of extinctions and periodic cataclysms, only the "tough" ecosystems remain. However, technologically advanced, industrial human societies create novel perturbations outside the realm of conditions experienced in the history of most terrestrial ecosystems. Humans transport and introduce new species, alter the patterns of disturbance and baseline conditions, and change the configurations of landscapes.

The effect on the environment and the biota of industrial societies at high population densities has been profound. Inasmuch as five of the great extinction events may have been due to asteroid impacts with the Earth, the actions of industrial human society have been likened to a "sixth asteroid"—because the high level of species extinctions over the past few hundred years has been caused mainly by human alteration of the planet.

Change is an essential part of nature. Few, if any, observations of ecological systems have displayed long-term constancy. Change in one part of an ecosystem potentially transmits change to another. Animals can alter the vegetation; altered vegetation changes the quality of habitat for animals. This feedback modifies the diversity of animals, which can in turn alter the animals' impact on the environment, and so on and so on.

Human activities, along with variations in the natural environment, inevitably initiate global and regional change. We need to understand where these changes will take us. A basic question is whether or not the magnitude of these changes—iterated through ecosystem feedback—decreases, as do the ripples when a rock is thrown in a pond. Or whether the ripples are amplified into waves, small changes becoming larger over time. We need to know which is the case.

2 The Big Woodpecker That Was Too Picky

On the previous evening my father and I, with two companions, had entered the Singer Preserve, near Tallulah. This area at the time was a great virgin bottomland forest. We were in the quest of America's rarest bird, a species that few living ornithologists had ever seen except as a museum specimen . . . After several unsuccessful attempts to see this great woodpecker in the Singer Preserve in the summer of 1934, I was still trying on the Christmas day mentioned above. My companions and I were out at daybreak . . . A slow drizzling rain that began to fall did not seem to better our prospects, but suddenly, in the distance through the great wood, a telltale sound reached our ears. Approaching cautiously in the direction indicated by the calls, we soon beheld not one but four Ivory-bills feeding on a tall dead snag! There were two males and two females, which with their powerful bills were proceeding to demolish the bark on this dead tree, in search, no doubt, for flat-headed beetles, or "betsy-bugs" . . .

It is possible that no future generation of Americans will be able to spend a Christmas morning, watching four Ivory-billed Woodpeckers go about their daily routine amid huge redgums whose diameters are greater than the distance a man can stretch his arms. I wonder what natural beauties we shall have, aside from the mountains and the sky, a hundred years from now!—George H. Lowery, Jr. (*Louisiana Birds*, 1955)

Sadly, George Lowery's concerns proved to be well founded. It is highly likely that the ivory-billed woodpecker (*Campephillus principalis*) no longer exists. It was a striking creature in the floodplain forests of the great southern rivers of the United States. An extremely large black-and-white woodpecker with a red crest and white bill, it was never common. Very tame, it made an easy target for commercial hunters, who shot the animals to sell as specimens for wealthy gentleman collectors. This practice certainly decimated the population, but the bird's vulnerability ultimately stemmed from its need for extensive areas of mature floodplain forest.

The highest regional density ever recorded for ivory-billed woodpeckers was only about one pair per 36 square miles (50 km²).[1] The reason for this low density appears to be the bird's specialized feeding habits and the scarcity of its feeding stations. As Lowery and others observed, the bird got its food by pecking away the loose rotting bark from large standing dead trees (Figure 1). Once a woodpecker had removed all the bark from a dead tree, it needed another standing dead tree with loose bark and available insects as a replacement feeding location.

How many large dead snags does a floodplain forest contain? At what rate are these replaced, to provide an Ivory-billed woodpecker with a steady supply of places to feed?

Unfortunately, such snags appear rarely. A suitable dead tree occurs 200 to 400 years after the tree's birth as a seedling, its growth to maturity, and its death. Only when a large, mature tree dies while still standing will the resultant snag be usable by a feeding specialist such as the ivory-billed woodpecker—and then only for a year or two.

This consumption of a rare element of the forest illustrates how the mature forest is a dynamically changing and continually regenerating mosaic. If there is enough of it, this forest mosaic can regenerate the feeding habitat of the ivory-billed woodpecker as a sustained and renewable resource. Because of the slow rate at which feeding sites are produced, a large area of land is needed for enough dead trees to allow a pair of birds to raise a brood of young. This requirement is at the heart of the woodpecker's eventual demise. The bird's rarity requires an understanding of forest dynamics.

Vegetation looks like broccoli when viewed from above at some level—it is clumped and patchy.[2] A forest canopy is a mosaic of tree

Ivory-billed Woodpecker.

1 Male 2 & 3 Female

Figure 1. Three ivory-billed woodpeckers (Campephillus principalis) *feeding on beetles by flaking the loose bark from a dead tree. From J. J. Audubon,* The Birds of America, *vol. 4 (Philadelphia: J. B. Chevalier, 1840).*

Figure 2. *The Australian subtropical rain forest in Lamington National Park (Queensland, Australia), showing the forest canopy as a mosaic of trees.*

crowns, interspersed with holes in the canopy (Figure 2). The holes, which forest ecologists call gaps, result from the violent deaths of large trees. A rich body of theory about the ever-changing nature of landscape mosaics is based on these forest gaps and their changes over time. The central concept is that the elements constituting the forest mosaic change cyclically; a canopy gap is one phase of this cycle.

To visualize the forest cycle, imagine that you are standing in a mature forest viewing a small plot of land dominated by a single large tree. The large tree shades the ground, thereby reducing the survival of smaller trees and seedlings below. A few smaller trees of shade-tolerant species survive under the large tree, but they are spindly and their growth is suppressed. The canopy tree dominates the available resources (light, water, and nutrients).

If it were possible to speed up time, how would the forest change from this starting point? When the dominant tree dies, the forest floor (where previously a young tree had little chance of a survival) abruptly becomes a nursery for small seedlings and saplings. With adequate light and other resources, hundreds of seeds germinate. The resultant small seedlings survive and begin to grow toward the canopy. The small trees that struggled under the dominance of the now-dead canopy tree begin a race with the seedlings. They compete with one another to become the tallest. Some lose to more vigorously growing competitors. Eventually, one tree becomes the locally dominant canopy tree and begins to eliminate the others. The site returns to a state much like the starting point and closes the mosaic cycle. A large tree again dominates the site. When it dies, the cycle begins again.[3]

Forest ecologists have generated much of the theory about land-scapes as dynamic mosaics. A. S. Watt, a forester, developed his doctoral research in 1925 on the pattern of vegetation in a mature European beech (*Fagus sylvatica*) forest in southern England.[4] His surveys led him to develop an important concept in ecology: that uniform processes are responsible for the heterogeneous patterns seen in all vegetation. Over the next two decades, Watt perfected his ideas by studying a wide range of British vegetation types. In 1947, by then a recognized leader in plant ecology, he gave the presidential address before the British Ecological Society and extended these concepts to a range of other systems (bogs, sparse grasslands, alpine communities, heaths).[5] Watt's address, a classic paper in modern ecology, presented two sorts of examples to support his theory.

First, he developed cases in which he systematized a seemingly helter-skelter spatial pattern of vegetation mosaic to elucidate the underlying processes causing the pattern. Interpretation required the observer to deduce the appropriate ordering of the patterns in the vegetation. Once this botanical jigsaw puzzle had been solved and the correct order determined, the process causing the pattern was revealed.

For example, the heterogeneity of the mature European beech forest at the core of his doctoral work (Figure 3) was the product of underlying cyclic dynamics. The pattern of the forest resulted from the fact that different parts of the forest floor were in different parts of the underlying cycle. Watt documented several other such explanations of heterogeneity in very different British ecosystems: the patchy surface of a windswept sparse grassland on sandy soil in Breckland, near the center of Norfolk; clumps of different kinds of shrubs in British heath-lands; the hummocks and hollows of moors. In all of these examples, the size of the dominant plants and the cycle of the colonizing, growing, maturing, and senescing of individual plants created a mosaic pattern across the landscape.

Second, Watt discussed cases in which the cyclic processes seen in the examples above were naturally synchronized to some degree, usually by environmental conditions. No longer helter-skelter, the synchro-

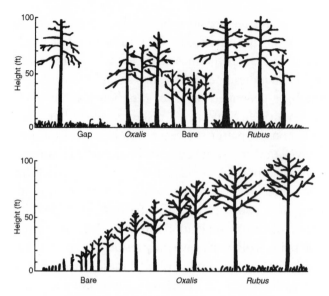

Figure 3. Mosaic dynamics of mature European beech (Fagus sylvatica) forests in England. Above, Cross-section through a mature European beech forest. Notice the variation in the sizes and arrangement of the trees, as well as in the ground vegetation (Rubus, Oxalis). Below, Rearrangement of the variation in vegetation as different stages of the gap replacement process. From A. S. Watt. Pattern and process in the plant community, Journal of Ecology 35 (1947):1–22.

nization resulted in vegetation patterns that moved in a predictable way over the landscape. Watt's examples were from vegetation in relatively adverse locations, notably from high in the Cairngorm Mountains in Scotland. They involved the same species—shrubs such as the common heather (*Calluna vulgaris*) and low-statured herbs and grasses. These cases are compelling because there is no need to infer the ordering of the patterns of the vegetation to understand the underlying processes that generate the patterns; the vegetation literally arranges itself.

Since Watt's 1947 paper, vegetation of larger stature has demonstrated this same tendency to be arranged in regular patterns. Two examples of these so-called self-organizing systems are "tiger bush" and "fir waves." Both reinforce Watt's interpretation of the cyclic processes of growth, death, and rebuilding as the generators of vegetation pattern.

Tiger bush occurs in regions of arid or semiarid climates with infrequent but highly intense rains. In mapping surveys of the Sahara Desert after World War II, extensive aerial observations revealed parallel stripes of alternating vegetated and nonvegetated zones.[6] These patterns resemble the stripes of a tiger; hence the English name tiger bush, or the French *brousse tigrée.*" This strange phenomenon (Figure 4) occurs in Africa, North America, and Australia.[7]

Tiger bush is found on flat areas with gentle, regular slopes where the soils are not very permeable. When it rains, water does not penetrate the soils and often produces runoff that flows over the surface in sheets. Where there is vegetation, the soil structure is altered by the presence of roots and leaf litter and allows rainfall and runoff to soak into the soil. This moisture irrigates the vegetated band that forms the tiger "stripe." The plants at the front of the tiger stripe have first access to water, then the stripes migrate slowly up the slope.

For example, in the Mapimí Biosphere Reserve in the Chihuahuan Desert of Mexico, the stripes have widths between 65 and 230 feet (20–70 m) and lengths of 300 to 1,000 feet (100–300 m). The areas covered by such patterns can be extensive. For example, tiger bush covers 32 percent of a 664-square-mile area (172,000 ha) mapped in Mapimí.[8] Within each stripe are five relatively distinctive zones. The major woody species forming the vegetation stripes (in Mexico, *Prosopis glandulosa* and *Flourensia cernua*) regenerate better on the "first-to-drink" or upslope side of the stripe. Plants colonize this side during years of relatively high rainfall. Plants die on the "last-to-drink" downslope side of the stripe in relatively dry years.[9] Thus, the movement pattern of the stripes varies, but generally it reflects the regional climate variability.

Basically, the tiger stripes arise from a feedback interaction of the vegetation, the soil, and the climate. The presence of vegetation increases the amount of dead plant material in the soil, which in turn significantly increases the rate at which water penetrates the soil in the zone above the vegetation stripe. The soil area ahead of the vegetation gets more water and becomes more hospitable to plant growth. As a result, the vegetation crawls slowly up the gentle slope of the desert, propelled by colonization of the now-moister zone.

Our immediate suspicion is that some organizing force (such as hedgerows, tree plantations, or stripes of willows along a river) produces regular patterns of vegetation. But tiger stripe vegetation sur-

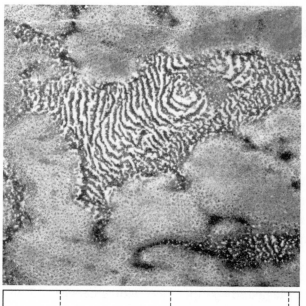

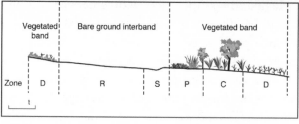

Figure 4. Above, *Tiger bush, or brousse tigrée, photographed from the air in Niger. The dark stripes are trees and shrubs. The photograph covers approximately 2 miles (3 km) on each side.* Below, *Cross-section of a brousse tigrée showing different vegetation zones. The soil in zone R (the run-off zone) has a surface layer that is difficult for the water to penetrate. The runoff water penetrates the soil when it reaches zone S (sedimentation zone) because the organic matter produced by the vegetated band loosens the texture of the soil. This penetrating water irrigates the vegetated band. Zone P (pioneer front zone) is the youngest vegetation, zone C (central zone) has the tallest vegetation, and in zone D (degraded zone) the vegetation is in a state of decline. The vegetation moves from right to left in the diagram because the upslope areas have access to rainwater before the downslope parts. From J. M. Thiéry, J.-M. D'Herbès, and C. Valentin, A model simulating the genesis of banded vegetation patterns in Niger,* Journal of Ecology 83 (1995):497–507.

prises intuition: brousse tigrée in Niger and tiger bush in Mexico orga-
nize themselves into patterns, just as Watt's examples from the Cairngorm Mountains do.

Consideration of a surprising geometric regularity in the vegetation
of the Sahara Desert seems a leap from understanding the habitat of a
large woodpecker in the floodplains of the southern United States.
However, the brousse tigrée is a clear illustration that vegetation at fine
spatial scales changes cyclically. The bizarre geometry of the tiger
stripes reveals the underlying dynamic cycle of the vegetation. A next
example moves closer to the forests of the ivory-billed woodpecker.

The second instance of self-organizing patterns is a naturally forming landscape geometry described as "fir waves" or "wave regeneration."[10] Crescent-shaped bands (or, in some cases, parallel stripes) of
dead trees are arrayed regularly across mountainsides. Their appearance in the high-altitude forest of Japan has been described thus: "In
the dark green of the gentle southwest slope of Mt. Shimagaree with
subalpine coniferous forest, several whitish stripes horizontally run-
ning in parallel with each other can be seen in a distant view so dis-
tinctly that the mountain has been named 'mountain with dead trees
stripes.'"[11] Fir waves have been observed in high-altitude conifer
forests at several locations around the world.

In New England, the stripes of dead trees represent a slowly moving
wave of synchronized mortality of balsam fir (*Abies balsamea*) trees.
This wave of death is followed by a trailing wave of regenerating and
growing trees, which in turn is followed by another wave of death. The
waves travel as if propelled by the prevailing winds. The time needed
for a wave to travel through an entire cycle is on the order of fifty-five
years, which means that the movements of the waves are from 3 to 10
feet (1–3 m) a year.[12] Faster-moving waves are found either on level
ground or moving downhill.

Fir waves strongly resemble Watt's inferred pattern of gap regeneration. His diagram of forests generating and repairing canopy gaps
caused by the death of large trees (Figure 3) looks exactly like a cross-
section through a fir wave found in the balsam fir forest in Maine (Figure 5). Internal ecological processes interact with environmental factors, giving fir waves a capacity for self-organization. These are special
cases in which the underlying interactions discussed by Watt for forest
are directly observable on the landscape.

Figure 5. Balsam fir (Abies balsamea) *waves in the mountains of Maine. Above, The white stripes are the exposed trunks of dead trees (photo by Meg Ounsworth). Below, A cross-section through a fir wave. From D. G. Sprugel, Dynamic structure of wave-generated* Abies balsamea *forests in the northeastern United States,* Journal of Ecology *64 (1976):889–911.*

Because Watt so strongly joined "processes" (death of trees, growth of trees, germination of seedlings) to "patterns" (tiger stripes or fir waves), his ideas are often captured in the phrase "pattern and process." That processes can cause patterns is particularly obvious in self-organizing vegetation, such as fir wave and tiger bush. The pattern of vegetation and its changes also can cause processes to change. Pattern and process are the yin and yang of vegetation dynamics; each interlocks with and changes the other.

The overhead architecture of forest canopies has strong implications for the processes operating inside forests over time. Trees can grow large enough to alter their own local environment and that of any subordinate trees. How strongly this effect occurs depends on the species, shapes, and sizes of trees involved. The environmental condition on the forest floor profoundly influences the regeneration success of different species of trees.[13]

Feedback from the canopy tree to the local microenvironment and subsequently to seedling and sapling regeneration influences which tree will become the next canopy tree that dominates the site.[14] These interactions drive the working of a forest and determine its nature when mature.

The environmental alteration of the forest floor caused by a canopy tree is most easily observed in terms of light.[15] Walk through a forest on a sunny day. Shadows and sun flecks cover the forest floor; the light is not homogeneous owing to the plants overhead. Along with light, other potentially important tree-environment interactions take place. Plants too can alter their local environment with respect to soil moisture, nutrient availability, soil temperature, and the amount of rainfall coming through the forest canopy. One of the most readily observed effects on the recovery of the canopy gap is that caused by the death of a tree.

When small trees die, the surrounding trees grow lateral branches and the site is relatively unchanged. With the death of a somewhat larger tree, the saplings growing underneath the tree may increase in size, and one of them may increase enough to begin to dominate the site. Small-tree gaps may not allow enough sunlight to reach the forest floor to trigger germination of the seeds on the shaded forest floor. Depending on a variety of factors (the latitude, which determines the average overhead angle of the sun; the height of the canopy; and the size of the gap), some size of tree is sufficiently large that, when it dies, light and other environmental factors on the forest floor change radically. Such tree falls are gaps in the usual sense and promote a competition among seedlings and existing small trees to control the location. Very large tree falls (or multiple tree falls) often favor the germination success of trees that disperse well, need high levels of sunlight, and grow rapidly. Certain existing tree species require large and highly disturbed sites. Some of the more important commercial tree species, notably various species of pines (*Pinus* spp.), are in this category. The standard treatment for promoting the next generation is clearing large areas of land ("clear-cutting") and sloughing or ripping the earth to expose mineral soil (called site preparation by foresters).

The fall of a tree and its attendant destruction have also been called the chablis, in part to sharpen the focus on the patterns of heterogeneity produced.[16] "Chablis" actually comes from an old French word that means an opening in a forest. The variety of grape used to produce chablis wine may derive from the gap-colonizing habit of the original vines.[17] The chablis has several important parts. The area beneath the crown of a fallen large tree undergoes a major change in light level at the forest floor. If the tree is uprooted, bare soil may also be exposed. The interwoven roots rip up a heap of soil as they are levered from the ground. A pit, which may be filled with standing water, replaces the space once filled by the heaved-over root ball (Figure 6).

In the area immediately adjacent to this root zone, the fall of the tree can have relatively little effect. Unless the fallen tree happens to strike another relatively large tree and topple it as well, the light levels at this adjacent site are relatively unchanged. However, the fallen tree lies on the ground and creates a microsite that is different from the rest of the forest floor. Some species of plants require elevated sites to regenerate. Farther away from the point of tree fall, in the area where the crown of the tree has landed, considerable destruction of the other trees living in that location may take place, as well as exposure of mineral soil and increased levels of sunlight. In this location, species that demand high light levels have an opportunity to grow into the canopy in the otherwise shady forest.

The death of a single tree kaleidoscopically produces a micromosaic on the forest floor, with the different pieces favorable to the regeneration of different species. Tropical rain forests, for example, may thereby maintain high species diversity. Long after a chablis event in rain forests (and in other forests as well), one can see the evidence of its occurrence in the patterns of the trees. Sometimes a straight row of trees of the same species, all about the same size, will be found. These are trees that regenerated on the trunk of a long-downed tree. Often they will be on "stilt" roots that grow on top of the downed tree trunk and send roots to the ground. When the tree trunk decomposes, the new trees are perched on the stilts that once wrapped around the downed log. Giant trees that are fast-growing, light-demanding species may emerge from the shady forest floor to occupy the canopy; these individuals are likely to have originated in the zone of destruction caused by the fall of a massive tree crown.

Just as the biology of the species determines regeneration success in

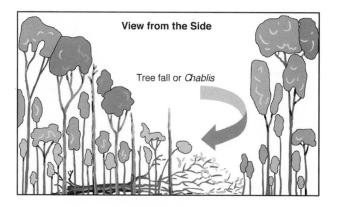

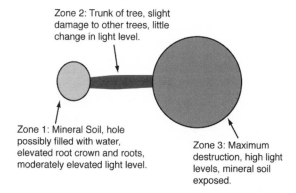

Figure 6. A tree fall, or chablis. *In a mature forest it produces a complex pattern of altered microenvironments. Thus, the death of a tree of one species leads to regeneration opportunities for several species with very different regeneration needs. From F. Hallé, R. A. A. Oldeman, and P. B. Tomlinson,* Tropical Trees and Forests *(Heidelberg: Springer-Verlag, 1978); and R. A. A. Oldeman,* Forests: Elements of Silvology *(Berlin: Springer-Verlag, 1991).*

gaps of various sizes and in different parts of the chablis, species attributes also determine the size and nature of gaps created by canopy trees. For example, the yellow poplar or tulip tree (*Liriodendron tulipifera*) of eastern North America almost always dies violently. One often finds blown-over trees with their leaves still fresh and green; almost never are dead *Liriodendron* trees found standing. Violent death arises from stress and old age weakening a tall tree's root system so that it topples with a strong wind. Other species grow differently and respond to stress differently; they die slowly and leave the standing dead snags with flaking bark that provide food for the ivory-billed woodpecker.

If mode of regeneration and mode of death are under biological control, then evolution might favor joining these features in a way that would promote species survival. An extreme example of such a conjunction is seen in the so-called suicide tree.[18] *Tachigalia versicolor* is a very large, highly branched canopy tree species that is found in evergreen and partially deciduous lowland forest in Panama, parts of Costa Rica, and Colombia. A tree blooms only once in its lifetime and then dies within a year after releasing a wind-dispersed fruit. Several, but usually not all, of the trees in a given area bloom, pollinate one another's flowers, and die at the same time. The *Tachigalia* seeds require a gap opening in the forest canopy for successful regeneration and subsequent growth and maturation. The death and fall of the parent produces a gap that greatly increases the opportunity for seedlings to grow to adulthood. Saplings of this species are often found growing in openings created by the fall of the dead adult *Tachigalia* that likely produced them.

What one sees in these suicide trees is a successful battle tactic of coupling the mode-of-birth and mode-of-death attributes of a tree. The violent death of the mother tree creates regeneration sites for its seedlings. The species creates in death the conditions needed for successful growth of the next generation. *Tachagalia* is an extreme case, but certainly other trees promote in the manner of their deaths the conditions needed for regeneration of their species.

Plants battling for ownership of space in the forest mosaic are locked in a complex battle full of enemies, allies, and intrigue. Thomas Hardy insightfully captured this dynamism in the following verse, excerpted from "In a Wood":[19]

> Heart-halt and spirit lame,
> City-opprest,
> Unto this wood I came
> as to a nest;
> Dreaming that sylvan peace
> offered the harrowed ease—
> Nature a soft release
> from men's unrest.

But, having entered in,
Great growths and small
Show them to be men akin—
Combatants all!
Sycamore shoulders oak,
Bines the slim sapling yoke,
Ivy-spun altars choke
elms stout and tall.

How do trees promote themselves and suppress their enemies? What are the broad strategies and the tactics of close infighting that are used by combatants in the war of trees?

A simple classification of the strategies can be developed from two questions: Does a tree species require a gap to regenerate? and Does a tree species generate a gap when it dies? The combinations of answers identify four broad strategies in a tree's battle to dominate the forest canopy. For *Tachigalia,* the answers to both questions are affirmative, and the species is what is known as a Role-1 species of tree. The four roles are as follows:

	Require Gap to Regenerate?	Generate Gap?
Role 1	Yes	Yes
Role 2	No	Yes
Role 3	Yes	No
Role 4	No	No

These strategic roles of trees are intentionally simple; one could easily make the categorization more complicated. However, even these roles can help us to gain insight into the potential complexities of interactions in a forest.[20]

Successful species in different categories influence the other species on a small patch in different ways. When a large, mature, Role 1 tree dies, it would be expected to create a large gap that would encourage its own regeneration, as well as that of all the other Role 1 species. In addition, it would encourage the Role 3 species in the community that also need gaps to regenerate. Trees in the other categories play out their strategies to influence their own regeneration and that of other species. The interplay creates a complex web of species interactions (Figure 7) on the mosaic battlefields of the forest landscape. Even with

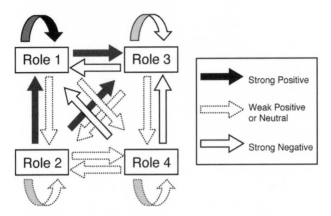

*Figure 7. The complex relationships among trees of four simple
ecological roles. Existing alongside the competition among
individual trees are other interactions that could be classified as
predation or mutualism.*

just four simple strategic roles, a forest has a complexity of interactions
comparable to the royal entourage in the court of a medieval Italian
castle.

The overall pattern of interactions among trees of different roles includes also a considerable mutualism (where two species reciprocally
help each other) along with predatory interactions (where one species
gains from another without reciprocation). For example, a Role 1
species creates gaps and has a positive effect on Role 3 species, which
need these gaps. Role 3 species do not make gaps and thus have a negative effect on Role 1 species. This Role 1–Role 3 interaction is positive
on the one side but negative on the other. In classic ecology, this asym-
metry resembles the interaction between a prey and its predator, or a
host and its parasite. The interactions among plants that involve with-
holding or losing space are as complex as the webs of interactions of
competition and predation in animal communities.

Given this complexity, how should a forest change over time? It
seems that this question has two answers, depending on whether we
consider the changes on the scale of a gap or on the scale of the entire
forest mosaic.

The answer is most straightforward on the gap scale. Considering
the forest gap-replacement cycle, the total living mass of trees (bio-

mass) at a location drops abruptly with the death of the dominant large tree. The biomass then slowly increases as the regenerating trees compete and the new winner emerges and fills the gap. Thus, a graph of biomass over the centuries for multiple cycles of gap generation and gap filling should have a distinctive "saw-toothed" shape. The teeth in this graph have sharp points at times when a large canopy tree has died. The distances between the teeth reflect how long trees live and how fast they grow to dominate a canopy gap. The saw-toothed curve is the pattern one would expect for the small area controlled by a canopy tree, an area perhaps 60 feet in diameter (about 1/40 ha).

What is the expected change over a large forest? A forest landscape's biomass dynamic is the sum of the changes of all the pieces of the landscape mosaic. If a landscape were devegetated by a single event such as a clear-cutting or a wildfire, then all the small patches of the former forested landscape would be treeless. The living biomass of the entire landscape would be near zero. As trees began to grow on all the patches, the number of living plants on the landscape would increase. After the passage of time, all the patches of the landscape mosaic would be covered by large trees. With every patch of the forest occupied by a large tree, the living biomass would be at a maximum. However, this maximum would not be sustainable. Large mature trees eventually become senescent and die. With the death of large trees covering the landscape, the total amount of biomass on the landscape would drop. Of course, the geriatric trees would not all die during the same year. Growth in the canopy gaps would eventually average out the losses from death, and the amount of living biomass on the forest floor would be in a state of balance. The landscape would be covered with a mosaic of patches of all ages, representing all of the stages of the underlying gap dynamics cycle.

How does the mature forest (or "virgin forest" or "climax forest") look? It should have patches with all stages of the gap-replacement cycle, and the proportions of each should reflect the duration of the different gap-replacement stages. The ecologists F. H. Bormann and G. E. Likens depicted a mature forest landscape as a dynamic mosaic of changing patches: "The structure of the ecosystem would range from openings to all degrees of stratification, with dead trees concentrated on the forest floor in areas of recent disturbance. The forest stand would be considered all-aged and would contain a representation of most species, including some early-successional species, on a continuing basis."[21] T. C. Whitmore expected this pattern and underlying process to work for all forests and asserted that the "forests of the world are fundamentally similar, despite great differences in structural complexity and floristic richness, because processes of forest succession and many of the autecological properties of tree species, worked out long ago in the north temperate region, are cosmopolitan. There is a basic similarity of patterns in space and time because the same processes are at work."[22]

The processes to which Whitmore refers are gap-replacement processes—the processes that produce dynamic forest mosaics. Indeed, the occurrence of such patterns has been documented for several kinds of forests.[23]

There are some exceptions. The scale of the mosaics in many natural forests is somewhat larger than one would expect from gap filling of single-tree gaps, indicating the importance of other phenomena that cause multiple-tree replacements.[24] In forests of very high latitudes, the sun is at an extremely low angle; a gap caused by a single tree is unlikely to create a canopy hole large enough to allow direct sunlight to hit the forest floor.[25] Some of the relatively longer records of changes in forest structure and composition indicate that the forest composition fluctuates.[26] Certain species seem to have periods of relatively weak recruitment of seedlings to replace large trees, but then undergo strong recruitment in other periods.[27]

The mature forests of the world are not collections of very large trees. Rather, they contain trees of all sizes in mosaic. They are not a primeval constant. They are dynamic, ever changing, and heterogeneous.

The sad demise of the ivory-billed woodpecker reveals much about the dynamics of natural vegetation. From the forest in which it occurred, the bird appears to have required a short-lived but slowly generated portion of the gap-replacement cycle. Its feeding habits required large, standing-dead trees with loose bark and insects underneath, a condition that is not always generated in the cyclic change in a forest; some trees, because of winds and other factors, fall over while still alive.[28]

The woodpecker "used up" its scant resource rapidly because of its habit of only removing the loose bark of the trees to search for insects. A large tract of mature forests would be required to supply enough forest mosaic elements to satisfy the bird's needs. The clearing for agriculture of the floodplain forests of the U.S. South spelled trouble for a large animal that required a significant amount of mature forest to meet its highly specialized feeding requirements. And, certainly, the shooting of this rather tame bird by professional hunters was a contributing factor.

3 The Black-Headed Bird Named Whitehead

Blew strong, yet the ship still Laying too, now for the first time saw some of the Birds call'd Penguins by the southern navigators; they seem much of the size and not unlike *Alca pica* but are easily known by streaks upon their faces and their remarkably shrill cry different from any sea bird I am acquainted with.—Joseph Banks, HMS *Endeavour* Journal, January 7, 1769

Smooth water and fair wind: many Seals and Penguins about the ship, the latter leaping out of the water and diving instantly so that a person unus'd to them might easily be deceiv'd and take them for fish; plenty also of Albatrosses and whales blowing very near the ship. We were now too sure that we had miss'd Fauklands Islands and probably were to the Westward of them.—Joseph Banks, HMS *Endeavour* Journal, January 8, 1769

Sailing the far southern Atlantic in a solitary tall ship, heading for the roaring gales of the Great Southern Ocean, and being aware of the imminence of the difficult passage through the Straits of Magellan would unnerve any sailor. Joseph Banks, looking out at the sea from the rail of the *Endeavour,* would have realized that he was in just such a situation. How could he know? By viewing large numbers of a bird that he had never seen until the day before, but that he immediately recognized. The birds were penguins.

Figure 8. The rockhopper penguin (Eudyptes chrysocome) *of the Southern Hemisphere. From R. Lydekker,* The New Natural History, *vol. 4 (New York: Merrill and Baker, 1890).*

Penguins (Figure 8) come to our minds, as they did to Banks's, as the creatures most typical of Antarctica. The seventeen living species are all birds of the Southern Hemisphere, as were the thirty-two known species of extinct penguins.[1] Penguins have black backs and white bellies. All are flightless because their wings are modified to serve as flippers. Their legs are toward the rear tip of their torpedo-like bodies, an adaptation to an aquatic life. This strongly posterior leg position makes the birds totter when they walk. Clumsy on land, the birds quite literally fly through the water. The ability of the animals to swim gracefully at surprising speeds makes them popular displays at large public aquariums.

Penguins feed at sea in the southern parts of the Pacific, Atlantic, and Indian oceans. While associated with icy scenes from the Antarctic, they are also found in temperate and tropical waters—as far north as the Galápagos Islands (Galápagos penguin, *Spheniscus mendiculus*). Penguins probably were first seen by Europeans in the Portuguese expedition of Bartholomeu Dias de Novaes in 1487–1488 traveling around the Cape of Good Hope in southern Africa. The Portuguese voyage of Vasco da Gama in 1497 produced the first written documentation of penguins.[2]

Their name derives from two Welsh words. *Pen* is Welsh for head, and *gwyn* means white. Some penguins have crests, eye decorations, and what might be described as bad haircuts. While certain species of penguins may have some white *on* their heads, no penguins have white heads. They are not named for their appearance at all; they are named after an island, a place far from where they are found. The "white head" that names penguins is a great, guano-whitened headland on an island near Newfoundland in the North Atlantic. It was a major breeding site for the now-extinct great auk, a black-backed and white-bellied diving bird slightly over 2 feet (70 cm) long. The island remains a major nesting ground for smaller auks and related birds. The photogenic little puffin (for example, the Atlantic puffin, *Fratercula arctica*) is probably the most familiar auk.

Great auks were found on White Head, or Penguin Island, in large numbers in the sixteenth century, when the island was given its name. On account of their rocky home, great auks (Figure 9) were called penguins long before Europeans misnamed the similar-looking but unrelated bird of the far southern oceans. In a sense, the great auk is still called a penguin, as indicated by its genus name, *Pinguinus*. The penguins of the Antarctic and southern seas were named for their resemblance to the auks of the North Atlantic, which were familiar to early explorers and sailors.

Joseph Banks's journal entries are part of his record during a great voyage of scientific discovery led by Captain James Cook on HMS *Endeavour*. On the adjacent days quoted above, Banks's observations of penguins capture two aspects of the way any species fits into its environment. One involves how the environment and its peregrinations af-

NORTHERN PENGUIN.

Published, July 13, 1799 by Harrison, Chase &c. No 11.38. Fleet Street

Figure 9. A bird labeled "northern penguin"; nowadays we would call it a great auk (Pinguinus impennis). From Naturalist Pocket Magazine *(London: Harrison, Chase, July 13, 1799).*

fect the pattern of distribution of a plant or animal species. The other involves knowing the internal interactions that structure ecosystems. Both aspects derive from definitions of the ecological term "niche." The niche of a species is the way the species fits into its environment and the way it interacts with other species.

On January 7, 1769, Banks recorded his initial sighting of penguins and noted that they looked and behaved like *Alca pica* (the razorbill), one of several species of black-and-white auks found in the North Atlantic ocean.[3] His comments relate to a phenomenon called ecological convergence. In convergence, unrelated species in different ecosystems are strikingly similar in appearance and function. Ecological convergence explains why the black-headed penguins were given the name of similar-looking black-headed birds that lived on a promontory of the faraway island called White Head.

Let us focus for the moment on Banks's January 8 observation and his commentary about penguins. Ships' navigators in this era could determine latitude and thus could know how far north or south they were. However, they could not measure longitude to determine their east or west position. With only a partial knowledge of their ship's location in uncharted waters, sailors focused on the surrounding sea for clues to their location. Banks saw the presence of penguins, along with whales and albatrosses, as possible evidence that HMS *Endeavour* was beyond its Falkland Islands landfall and was sailing into the perilous Southern Ocean. This association of a species (or set of species) with a particular array of environmental conditions is directly related to the ecological niche as it was initially conceived. Early in the twentieth century, "niche" was first formally defined.[4]

Even though the term is fairly recent, using knowledge of the needs of animals (or plants) to infer how to find them has been a major part of survival in hunting and gathering cultures. We can presume that human societies deep in prehistory valued individuals who could utilize such knowledge.

Today, certain people have an uncanny ability to locate individual animals of a rare species, even though the total numbers are small and the area to be searched is large. This ability, often a product of years of experience, is the mark of a successful hunter, naturalist guide, herb collector, or angler. It is also a point of pride among ecologists, wildlife scientists, and other field researchers. It was such a species-environment association that aroused Banks's suspicion that the HMS *Endeavour* had passed the Falkland Islands.

The California thrasher (*Toxostoma redivivum*), found in chaparral vegetation of that state, is one of several relatively similar and closely related birds that occupy different habitat types in the natural landscapes of California. In 1917 this bird caught the attention of Joseph Grinnell, a Californian naturalist-ornithologist interested in scientific explanation of the factors that control the distributions of animals.[5] In studying and observing the bird's behavior, Grinnell perfected an idea that he had hinted at earlier in a paper explaining the distribution of the chestnut-backed chickadee (*Poecile rufescens*), another bird of the western United States.[6] Based on his studies of the California thrasher, Grinnell defined the ecological niche of a species as the attributes that determine how it fits into different environments. Understanding the niche of an animal species would allow prediction of its potential geographic distribution.

The niche of the California thrasher, Grinnell believed, included the thrasher's foods, its preferences for certain types of vegetation structure, and other details that influenced where it could be found. Later, in 1928, he formally defined the niche as the "ultimate distributional unit, within which each species is held by its structural and instinctive limitations, these being subject only to exceedingly slow modification down through time."[7] At about this time, American and Russian botanists were developing essentially the same idea for plants; they referred to it as the "principle of species individuality."[8]

We have said that the aspects of a species necessary to predict its distribution and occurrence constitute the niche of the species. In applied ecology, the inverse—called the indicator species—is to use the presence of a species to diagnose environmental conditions.[9] Indicator species have a wide variety of applications from pasture management to prospecting for minerals.[10]

In most field guides for identifying plants and animals, description of a species' habitat often is essential information in separating one species from another. A bird's habitat might be characterized as moist forest or shrubby grassland. The overall condition of an animal's location is certainly related to the niche of the animal.[11] However, Grinnell's work indicates a broader concept—involving not only the habitat of an animal, but also its behavior and physiology. A species' niche provides a more formal basis for understanding how and where on the landscape a species might be found. Niche is concerned with what determines the geographic range of a species; habitat involves the setting where a species is usually located inside its geographic range.

We are in an era of large-scale alterations to landscapes and the environment. Prediction of the species that might be disadvantaged by these changes would be invaluable in managing the biotic diversity of the planet. Grinnell's autecological concept of species niche has the potential to provide this capability.[12] Indeed, in 1927 Grinnell recommended that the American Ornithologists' Union Checklist (recognized as the standard listing of the species and subspecies of all North American birds, along with information about where they are found) include the environmental limits of the ecological niche of species to provide a "system for designating birds' ranges."[13] In our present era of concern about global change and altered biodiversity, it is unfortunate that this lead has not been followed, and that seventy-five years of such information gathered across the face of a changing continent is not at our disposal.[14]

What difficulties would face ecologists attempting to develop the system that Grinnell recommended in 1927? It is not as easy as it initially appears to identify irrefutably the factors that control the distribution of a given species. Proper scientific procedure calls for experimentally manipulating the environment of species populations to determine which factors could cause a change in animal abundance. Unfortunately, *many* factors might affect any given animal and it would take many experiments to identify those that are important. Also, with the numerous kinds of plants and animals, it seems impossible to determine what factors control their geographic distributions without conducting experiments on them all. For this reason, studies of species niches often look for conditions that are consistently associated with a species' presence.

A significant complication in modern applications of Grinnell's concept of the niche springs from an interesting and somewhat philosophical consideration referred to as ecological scale. Scale is a crucial concept for understanding ecological systems. While its treatment can take on almost mystic proportions in some theoretical analyses, the concept is not foreign to everyday experience. A telephone question to a friend in Paris, "How do I get to your house?" would involve airline schedules and where to meet in a local airport. The same question to a friend in town would involve streets and turns after familiar landmarks.

The time and space scales of observations can complicate determination of the factors controlling distribution of plants and animals.

Two examples, the first involving space scale and the second involving time scale, will clarify the point.

The first example is from a classic study of the distribution in Missouri of a small flower, *Clematis fremontii*.[15] This particular clematis is a perennial herb with prominent veins on its heart-shaped leaves. Because of the thickened texture of the leaves, it is commonly called leatherleaf. The plant occurs only in Jefferson County, Missouri. There it is found in glades—rocky barrens that occur where there are outcroppings of dolomite on the drier south-facing and west-facing slopes of otherwise wooded ridges.[16] The most characteristic plants of the glades are a dark evergreen tree, the eastern red cedar (*Juniperus virginiana*), mixed with several prairie grasses such as bluestem (*Andropogon scorparius*). The distribution of leatherleaf has been characterized as having distinct patterns at five different levels (Figure 10):

- Geographic range: Leatherleaf is restricted to a small part of the state of Missouri, an area about 25 miles in diameter ($\sim$1,000 km^2). Some 1.5 million plants make up the total population.[17]
- Regional distribution: In the leatherleaf geographic range, the species is found only in the part of the range where eastern red cedar glades exist (most of these are found in four subregions, each perhaps 4 or 5 miles across (approximately 20 km^2). There are about 300,000 individuals to a subregion.
- Clusters of glades: The red cedar glades occur in clusters that have areas typically less than 100 acres ($\sim$40 ha). The leatherleaf populations of these sites contain about 30,000 individuals.
- Glades: Within a single glade, the leatherleaf plants clump in patches of a couple of acres (about 1 ha). Typically, these colonies have roughly 1,000 individuals.
- Aggregations within glades: Even in the clumps within glades, leatherleaf plants aggregate in small areas about 10 yards on a side (less than 0.1 ha). These aggregations comprise small populations of around 100 plants.

Leatherleaf is not randomly distributed. At almost any level of resolution, it occurs in clumps. Within these clumps are smaller clumps.

When we compare the appearance of plants from different spatial levels, we see regular differences among the leatherleaf plants from the different-sized clumps. Plants from different parts of the same glade

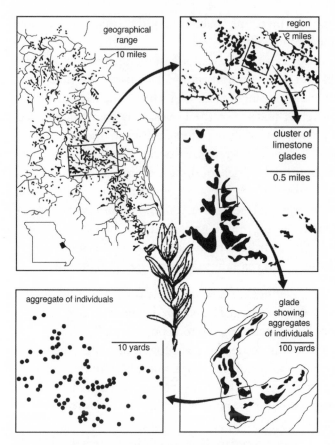

Figure 10. Five maps showing increasingly greater detail of the range of a small flower, Clematis fremontii, *in Missouri. The factors that predict distribution of the species at the regional level could be quite different from the predictive factors at a local level. From R. O. Erickson, The* Clematis fremontii *var.* riehlii *population in the Ozarks,* Annals of the Missouri Botanical Garden *23 (1945):413–461. With permission of the Missouri Botanical Garden.*

tend to have observable differences in their leaves. From one region to another, plants differ most in their flowers.

What causes clumps in the distribution of leatherleaf? From the regional level, the answer would be the presence of rocky, dolomitic outcroppings. If we consider the plants in a single glade, all the area is on dolomite outcroppings and the clumping in the glade must be caused by something else. The environmental factors that best predict the pattern of the species at one level are not necessarily those that predict the species at another level. Scale too has a role in predicting where a species will be found.

Scale-related considerations complicate the application of Grinnell's concept of niche, which is based on a total understanding of the species' needs. As a theoretical concept, niche would include the spatial scales of the factors that influence each species. The challenge is in execution of the theory. Two ecologists studying the very same leatherleaf plant at different space scales in Missouri could arrive at quite different factors controlling the distribution of the species. We are not likely ever to have data from multiple studies at multiple scales for the diverse species of plants and animals on our planet. In a practical sense, our predictions of the occurrence of a species from niche studies will have a level of uncertainty—no matter how much effort we focus on the problem.

Our second example involves time scales in ecology. If we could know the environmental factors essential to determine the location of a species, we would still have the problem of how these factors interact with the species in time. Again, this is a concept that we experience in our daily lives. We touch a moist finger to an iron to see if it is hot. Every winter the tabloid newspapers feature pictures of members of "polar bear clubs" plunging into icy holes cut in frozen lakes. We can quickly stick our hand into a very hot oven and remove it without harm. While these temperature extremes can be tolerated, they cannot be tolerated for very long. Similarly, the factors that interact to control geographic distributions of plants and animals have a time frame in which they cause responses; at other levels they do not.

Ian Woodward illustrated this concept when considering the responses of different plant processes to climate factors.[18] The temperature variation of climate has some periodicity: the 24-hour cycle of the day and the 365-day cycle of the year are the obvious examples. Other periodic changes occur in the climate, as well. For example, extreme long-term periodicities cycling over tens of thousands of years will be discussed in the next chapter.

Woodward conjectured that the expansion and contraction of the ranges of different types of plants (late successional trees, herbaceous perennials and annuals) should be controlled by longer frequencies of climatic variation. Because they have a long life and are large enough to exert a degree of local control over their environment, late successional trees would require climatic changes of multiple centuries to induce a contraction in their range. Such a contraction involves the death of established plants, a process that takes time. Because they are shorter lived, other types of plants should have a more rapid response than trees.

While the ranges of plant species are not influenced by monthly or daily variations in the climate, the periodic variation in temperature over these time intervals excites plant flowering and germination. The temperature variation felt when the sun goes behind a cloud, or a cool breeze touches your face, lasts minutes or seconds. These variations do not affect the ranges or the flowering of plants, but they are strongly involved with other important plant responses such as leaf drying or the rate of leaf photosynthesis.

Seemingly straightforward, these variations do run somewhat counter to popular, almost mystical, ideas that everything in the ecosystem is tied to everything else. Here is a signature verse for this view:

> Near or far,
> Hiddenly
> To each other linked are,
> That one canst not stir a flower
> Without troubling a star
> —(Francis Thompson, *The Mistress of Vision,* 1897)

How much does the stirring of a flower trouble a star? Not much at all. Thankfully, astronomy has not bogged down in the "flower-star interaction calibration problem." Similarly, progress in ecological science depends on determining which factors to ignore and which to include in understanding the interactions of parts of ecosystems. Knowledge of the appropriate scales in space and time for a given question is essential in this determination.

For mobile animals such as birds, Grinnell's niche concept implies that if one released a large number of individuals of several different species onto an empty landscape, they would sort themselves into various locations that matched their niche requirements. Species abundances would be proportional to the area containing suitable places for them. The species with inappropriate niches over much of the landscape would be rare and might move elsewhere or even perish. This expectation of the outcome is fairly simple, whereas the process could actually be rather complicated. However, the rules that produce patterns of commonness and abundance could be much more complex. A logical question is, "How is reality more complicated than a simple explanation?" The answer derives from an important procedure in landscape ecology, namely the use of "null models."

A null model is an explanation that predicts what should happen in a given situation. Usually, it is purposefully designed to be extremely simple and straightforward. When it does not work, then we are justified in suspecting that more complex effects are involved. Further investigations are needed to determine just what these effects might be. Null models assume that it is appropriate always to select the simplest (most parsimonious) explanation of a natural phenomenon.

The preference for a simple explanation over a complex one as a criterion for judging competing theories is associated with a thirteenth-century English monk, William of Occam, or Ockham (1284–1347). Occam was engaged in the medieval religious argument over whether God revealed knowledge to humankind or hid it. His principle, sometimes called Occam's Razor, stems from Aristotle's opinion that "entities should not be multiplied beyond what is necessary." In other words, always chose the simplest explanation—which is the basis for the transition from medieval to modern scientific procedures.

The null model which assumes that species sort onto the landscape according to their niches, without interacting with one another, is a model that in many cases is rather difficult to eliminate—this *despite* the complexities of defining precisely what a species niche is. For this reason, Grinnell's concept of the niche has a number of practical applications involving plant and animal distribution patterns in novel and changed environments.[19] These applications demonstrate the power of Grinnell's niche concept for practical management problems.

For example, the U.S. Bureau of Reclamation dredges the Colorado River for sand and silt that have eroded into the river and piles the spoils as sand islands. The islands are devoid of vegetation, but by pumping water from the river through a drip-irrigation system to small planted cottonwood (*Populus fremontii*) trees, one can create a rapidly growing forest. Since only trees that are planted and drip-irrigated become established in the initial stages of sand-island succession, the structure of vegetation can readily be controlled. A research group working on the niche relations of birds in Arizona manipulated plantings on these sand islands with the goal of creating an artificial habitat that would support a greater number of species per 250 acres (100 ha) than *any* native habitat in Arizona.[20] Using their decade-long studies of niches of Arizona birds, they accomplished their objective in three years. Earlier, they had developed measurements of vegetation structure as the niches of a long list of Arizona birds. These data were used to forecast the species found on the sand islands. The presence or absence of birds could be predicted correctly about 90 percent of the time in these experimentally manipulated habitats.[21]

The ability to interpret changes in habitat in terms of species presence for an entire assemblage of species has obvious implications for the management of biotic diversity. Changes in habitat structure will favor some species and disadvantage others. One major product of the Arizona study, from a habitat manager's point of view, was an interactive computer program capable of predicting which of the bird species in Arizona would be favored by different habitat management practices on the artificial sand islands.

Today the omnipresence of environmental change and human alteration in natural ecosystems has elevated the need to better understand what controls the abundance of plants and animals. For some species, survival in a changing world may be considerably aided by extension of Grinnell's ideas. New tools give us the ability to match a species' needs against environmental patterns. These tools include remote sensing of the Earth's surface using satellites to gain extensive baseline information on habitat patterns. Global positioning systems (GPS) derived from secret military intelligence are now available in unclassified form. GPS allows rapid location of the exact positions of ground observations. Powerful computers can then interweave layers of complex spatial data on the environment over large areas.

Grinnell's concept has found wide application in the management-oriented ecological sciences. New technologies are already augmenting these applications. Unfortunately, we are also changing some of the world's ecosystems at such a rapid rate that, even with technological advances, we may not have even rudimentary information on the requirements of certain rare species—before they are gone.

Grinnell's niche concept focused on a species' attributes and how these attributes determine geographic distribution. While Grinnell made it clear that it was vital to consider interactions with other animals (both predators and competitors), these were not his paramount factors. The Grinnellian niche is, in most of its aspects, a concept of individual organisms or species. The niche is an amalgam of the environmental factors involved in understanding why a given organism may occur in a given location.

When Joseph Grinnell studied the California thrasher and synthesized his observations to understand its geographic distribution, he was well aware of another of the bird's traits. The California thrasher is but one of a set of western thrashers with relatively similar body size and coloration. It is associated with and is adapted to the chaparral habitat. The other thrashers found in California and the western United States have similar affinities for shrubby habitats: Bendire's thrasher (*Toxostoma bendirei*) is found in deserts; Le Conte's thrasher (*T. lecontei*) in sagebrush and open deserts; the Crissal thrasher (*T. dorsale*) in thickly vegetated canyons. In the eastern United States only one species, the brown thrasher (*T. rufum*), occupies a variety of open shrubby habitats throughout the region.[22]

One could ask many questions about these thrashers: What determines the number of thrashers in western habitats? What is the pattern of differences among all these thrashers? Do thrashers play different roles in different shrubby habitats? What conditions could allow the species to be found together? These and an array of similar questions that could be asked about the California thrasher are associated with a different concept of the niche. A British ecologist, Charles Elton, developed this later niche concept with a fundamentally different intent.

Elton considered the interactions among species populations that produce regularities in ecological communities. He theorized that "the niche of an animal means its place in the biotic environment, its relations to food and enemies."[23] The features that interested Elton were the tendency for more small animals than large to be in a given location, for there to be a greater mass of herbivores than carnivores, and for unrelated creatures in very different locations to look and behave similarly (Figure 11). It is this latter phenomenon, ecological convergence, that allowed Joseph Banks to know penguins the first time he laid eyes on them. He knew that the penguins of the southern

*Figure 11. Convergence among African (left) and South American (right) rain-
forest animals. Pairs from top to bottom: pygmy hippopotamus* (Hexaprotodon
liberiensis) *and capybara* (Hydrochaeris hydrochaeris); *African chevrotain*
(Hyemoschus aquaticus) *and paca* (Agouti paca); *royal antelope*
(Neotragus pygmeus) *and agouti* (Dasyprocta leporina); *yellow-backed
duiker* (Cephalophus silvicultor) *and gray brocket* (Mazama
gouazoubira); *giant pangolin* (Manis gigantea) *and giant armadillo*
(Priodontes maximus). *From F. Bourlière, The comparative ecology of rain
forest mammals in Africa and tropical America: Some introductory remarks
(pp. 279–292), in B. J. Meggers, E. S. Ayensu, and W. D. Duckworth, eds.,*
Tropical Forest Ecosystems in Africa and South America: A
Comparative Review *(Washington, D.C.: Smithsonian Institution Press,
1973).*

oceans would look and behave like the familiar auks of the northern oceans.

Elton saw convergence as evidence of similar niches in geographically separated communities and reasoned that these similarities implied rules for the patterns of species niches.[24] If one could find analogous niches for species in communities in widely separated locations, then one also could think of communities as having "empty niches"— vacant ecological jobs waiting to be performed by an appropriate species, perhaps from some other part of the world. He emphasized interactions among the different populations of animals and plants. His initial concept stressed that "you are what you eat"; ecological community patterns could be understood from the feeding relations among species. In Elton's view, arctic foxes that feed on guillemot eggs and scavenge seal carcasses left behind by polar bears could be thought of as ecologically equivalent to the hyenas of semiarid Africa that eat ostrich eggs and the remains of lion kills.[25]

While Elton's idea of the niche included the species' place in the community with respect to shelter, competition with other species, and a range of other considerations, his conceptual focus was on the relation of the species to its food and its predators. That focus changed in 1932, when a brilliant experimental biologist, G. F. Gause, produced a clever set of laboratory experiments to understand competition between species.

Gause worked with three species of the microscopic protozoan genus, *Paramecium;* pairs of these species competed for food. The experiments used test-tube *Paramecium* populations fed by regular inoculations of yeast cells. The intent of these now-classic experiments was to test then-current mathematical theories for the interactions between populations of different species.[26] Gause's data strongly resembled (and in this sense, confirmed) the results predicted from theoretical analysis of dynamic equations for interacting populations.

The joining of mathematical descriptions of the growth, death, and competition of species with experimental results held the promise of moving population and community biology to a more rigorous, mathematically formal level—an exciting breakthrough for theoretical population ecology.

Gause observed that both of the non-coexisting paramecia fed in the top portion of test tubes, where each was able to survive in the absence of the other. Also, either competitor could coexist in the presence of a third species, which appeared to feed at the bottom of the test tubes. A top-of-the-test-tube feeder could live with a bottom-of-the-test-tube feeder, but two top feeders could not coexist. Gause associated his experimental results directly with Elton's ideas about the niche when he wrote, "As a result of competition two similar species scarcely ever occupy similar niches, but displace each other."[27] Today this is known as Gause's principle, or the competitive exclusion principle: no two species can occupy the same niche.[28]

Gause's experimental work narrowed the Eltonian niche in scope and shifted its emphasis from feeding interactions (which conduct one's interest to the relations between predators and their prey, or herbivores and their food) to competitive interactions. Internal factors that structure communities have remained central to the interest of researchers involved in competition studies and with other Eltonian issues such as the limit of similarities among species, or the degree to which different communities can be invaded by alien species owing to their empty niches. Both Grinnell and Elton pointed out that two species in widely separated communities can be thought of as occupying the same niche, but such occurrences would be more common in the more broadly defined Eltonian niche.[29]

Competitive interactions are relatively easy to demonstrate in laboratories by observing populations of small plants or animals in close quarters. However, demonstrability under natural conditions is a topic of debate.[30] Stable competitive interactions and/or competitive exclusion, the hallmarks of Gause's interpretation of the Eltonian niche, are more difficult to produce under field conditions. A rich array of circumstantial evidence has been used to support the concept that the structure of ecological communities results from competitive interaction.

One can conceive of ecological communities of plants and animals being shaped by the set of all the pairs of species having different relations.[31] Increase in a given species would augment its predators, decrease its prey, diminish the success of its competitors, and aid its mutualists.[32] In Chapter 2, we saw that various trees interacted in equivalently complex ways to occupy the space in a forest. Webs of positive and negative interactions theoretically could produce patterns in the abundance of plants and animals at a given location.[33] Rareness versus commonness, ratios of carnivores to herbivores, numbers of species, all are elements of what is called the structure of a community.[34] The mathematical analysis of formulations based on combinations of species interactions has been used to develop a theoretical foundation for community structure.[35]

Modern Eltonian theory is based on patterns in the shapes and sizes, the abundances and spatial distributions, of species in communities. These patterns include differences among individuals (degrees of similarity in shape and size) and species (patterns of distribution and abundance).[36] In some cases, the presence of one species appears to cause the elimination of another; in other cases, species in the presence of one another differ in ways consistent with having different niches. While there are exceptions, the modern Eltonian theory strongly emphasizes competition.

Theoretical treatments of competition as the structuring force inside ecological communities are topics of considerable debate among ecologists.[37] The issue is not whether competition occurs; we can demonstrate that it does. Rather, the issue is how competition can be used in truly predictive theories. Some ecologists consider competition to be a concept that has failed to develop into theories worthy of the name.[38] Others have noted the lack of reality of the equations that generate theories of community structure.[39] Still other ecologists endorse the importance of competition and are attempting to enrich the current theory by adding the effects of spatial pattern and unique aspects of the life history of plants and animals.

The result is a complex network of concepts that are internally consistent and, while close to the observations data, are often based on circular arguments.[40] Nevertheless, the interest in understanding niche theory and community pattern has generated a large observational database on the distribution and abundance of organisms. An increased appreciation for the dynamic and interactive nature of ecological systems has developed in the process.

A consistent observation in ecological communities is that if two species are considered with enough detail (or enough different niche factors), almost invariably ecological differences in the species will be found. There are two interpretations of this observation of separation of species niches in communities. The first is that communities are (or were) highly interactive biologically and that competition has worked at some level (such as behavior or evolution) to structure the communities so that there is little overlap in niche among species. This construction derives from considerations of competition and niche interactions. The second interpretation is that communities of animals (and plants) are relatively open. Niches are available for additional species, and species niches do not overlap very much. A significant difference in the two views involves whether or not the communities are so saturated that adding one further species would cause the loss of some other species. The concept of highly interactive communities is central to the development of community theory, but evidence suggests that real communities are a mixture of biologically noninteractive and interactive—with the noninteractive cases being more in evidence.[41]

Today, Grinnell's and Elton's niche concepts are applied at different levels with different objectives and largely with different explanations. The objectives of the Grinnellian approach are relatively straightforward. They involve determination of the attributes of a species and of the environmental factors controlling the range of the species. The geographic scale is large.

The Eltonian concepts are much more abstract. They involve community structure: ratios of sizes, numbers, and other patterns. The focus is on the ecological community, a more local assemblage of plants and animals. Thus, the space scale is relatively smaller than that in the Grinnellian applications. The processes that are considered may involve the evolution of the relevant species.

Communities tend to resemble one another physically in similar environments, even when they are composed of different species. For example, in Mediterranean climates in southern Africa, Chile, or Australia, the vegetation and plants resemble those in similar climates in France or Italy. But the species that make up the vegetation are not related taxonomically.

Species that are relatively unrelated taxonomically can have striking similarities in the structure of the vegetation as a whole, in the morphology of the component individual plants, and in the variation in the shapes and sizes of leaves. These patterns have been noted by ecologists and protoecologists from Theophrastus to Alexander von Humboldt to today's scholars.[42] The regularities imply underlying rules for the assemblage of communities.

What causes the underlying similarities of structure in ecological communities? Multiple explanations derive from multiple scales in geography, time, and organization. As an analogy to the question, "What causes ecological structure?" consider the question, "What makes a person happy?" Some people just appear to be happier than others; so one might expect intrinsic factors to have some role. Certainly conditions in the environment, physical living conditions, contacts with others, and a range of other extrinsic factors also are important. However, some happy people live in circumstances that appear difficult; some very unhappy people live in conditions that seem ideal to the outside observer. Subtle internal controls keep the happiness in those of us blessed with good mental health bounded somewhere between depression and lunacy.

Pursuing this analogy for ecological systems, populations may be internally controlled by physiological and behavioral attributes of the species, or by interactions with other populations (competition, predation, mutualism), or by external conditions in the environment.[43] Just as "What makes a person happy?" can only be answered in context, perhaps "What controls ecosystems?" is also context specific. It may be more appropriate to think of ecological systems as not being controlled at all by single factors. Rather, they are held within bounds by factors that under certain circumstances can alter the direction of change and thus confer a degree of stability to natural systems in a world whose environment is subject to constant change.

This chapter has treated two traditional views of the species niche— one involving the ways species interact with one another to produce patterns in ecosystem structure, the other concerning the ways the distribution of species is controlled by their attributes. Joseph Banks's observations of penguins in January 1769 combined both of these niche concepts: the Elton's niche with the ecological convergence of penguins and auks, and the Grinnell's niche with the penguin as an indicator of proximity to the Great Southern Ocean. Today we are looking at penguins as indicators of the effects of climate change on the Antarctic ice[44]—a modern analogue of Banks's use of penguins to infer the clime of the frigid Southern Ocean well over two centuries ago.

Beyond penguins, we face the same challenges Joseph Banks did: to recognize species requirements and to understand how ecosystems are put together. Both facets of the niche concept involve the understanding of how plants and animals are distributed both locally and geo-graphically. While the basic ideas were formalized early in the twenti-eth century, the terminology remains confused. Nevertheless, progress in anticipating where different species are found, and the challenging questions that are associated with predicting the structure of ecological communities, lend significance to the concepts associated with the niche and habitat of species. Grinnellian niche theory gives us a potential tool for managing habitats to support particular species. Eltonian theory, if it can be perfected and tested, could provide a basis for understanding the larger consequences of species loss and environmental change.

4 *The Rat That Hid Time in Its Nest*

The whole family went up in the hills to camp and gather acorns. Ashnat woman was always sick and lying down. She had two stones she warmed and laid over her eyes. They fed her only the acorn soup that stuck to the cooking stones. They gave her the stones to lick off. Her mother took care of her.

They all went to bed. One [person] thought, "I'm not going to sleep. I am going to stay awake and see what she is doing." This one listening could hear acorns rattling. This one who stayed awake was a visitor who had been with them from a neighboring family. She went home to her family and reported that she thought the sore-eyed Ashnat woman was the one who was stealing their acorns nightly.

[The neighbors] . . . then went over to visit the family of the sore-eyed one to see what they could learn. They found the house clear full of acorns. The children of the sore-eyed one said, "She does not go out because she has sore eyes." She was only making believe she had sore eyes. The visitors told her, "That's just the way you are going to be: stealing all the time from other people." She turned into Woodrat, and all her family likewise, and they scuttled away into the bushes.—Mary Ike (1940) telling the "Woodrat (Ashnat) Story," from Ashanamkarak, Siskiyou County, California

This Karok myth captures a distinguishing trait of several species of rodents called packrats. In the story, the packrats were transformed into prodigious gatherers of material stolen from others. In actuality, packrats busy themselves gathering bits and pieces that they use to build stick nests (usually with multiple escape routes), or middens. Because a packrat is equivalent to the food energy of about half a hamburger, it makes a worthwhile target for a wide array of predators. Midden building is a significant adaptation for a small animal that provides a nice hors d'oeuvre for a large predator such as a puma or bear, a light snack for a coyote, or a solid meal for a fox. Middens are important fortifications for these relatively vulnerable creatures.

In other stories, packrats swap sticks for other objects desirable for their middens. This "trading" occurs when a nest-building packrat carrying one article to its nest drops it to pick up another more interesting item—say Dad's car keys from a picnic table during the family's Arizona vacation. The past and current mythology of the creature is rife with stories about trading, stealing, and secreting objects.

Packrats (Figure 12) are big-eyed, large-eared, long-tailed rodents all in the genus *Neotoma.* They are found in a wide range of habitat types in the western United States and Canada. In all, twenty-one dif-

Figure 12. *Bushy-tailed woodrat* (Neotoma cinerea drumondii), *one of the twenty-one species of packrats found in North America. Packrats construct stick nests called middens, which can be preserved for thousands of years. Packrat middens preserve a record of the local vegetation when the animal was alive. From J. J. Audubon and J. Bachman,* The Quadrupeds of North America, *vol. 1 (New York: V. G. Audubon, 1854).*

ferent species are distributed over an area from the Arctic Circle in the Mackenzie District of Canada south into the tropics in Nicaragua (within 13 degrees of the equator), being most diverse in central Mexico. Habitats range from spruce-fir boreal forests in the far north to tropical forests in Central America and include a wide spectrum of vegetation across this expanse of latitude and climatic conditions.[1]

Many of the species prefer drier woody and shrubby ecosystems—deserts, chaparral, thorn scrubs, woodlands, and brushlands. Even the desert species of packrats require plant food that is at least 50 percent water. Also, they require shelter such as large rocks, shrubs, or trees.[2] Nearly all species build nests. They construct their dens using sticks and other plant fragments as well as bones, animal droppings, and in modern times various forms of litter (tinfoil, rings, and wristwatches of picnicking tourists). They urinate on their constructions to mark them and to glue the pieces together.

As packrat nests age, they undergo some rather remarkable transformations. After thousands of years in relatively arid environments, a substance called amberite forms from the packrat urine that covers the midden. A midden encased in amberite, called an indurated midden, resembles a brick of asphalt with a crackled surface.[3] The amberite covering is somewhat adhesive: indurated middens hang on cliff walls after the ledges that once supported them erode and collapse. In areas of softer sandstone, indurated middens are more resistant than the rock itself and stick to the surface of wind-eroded cliff faces.

The materials collected by packrats (and by a variety of other small mammals found in arid environments) and hidden in the middens can be used by ecologists to reconstruct vegetation at the site at the time the nest was constructed. Packrats build using materials that they collect from within 100 feet (30 m) of their middens. Therefore the contents of their nests provide quite specific information about the local vegetation. Because packrats are selective in the materials used in their constructions, interpretation of those contents usually involves knowing which species of packrat did the building.[4]

A midden is a detailed snapshot into the deep past. Paleoecologists study the ecology of past environments and can determine thousands of years into the past when packrat middens were built. Indurated middens preserve encased plant material—even intact pieces of plants—for tens of millennia.

Packrat middens reveal a deep history of vegetation change and yield some surprises about the past. In particular, this history differs from our expectations based on the writings of the great thinkers early in our human history. In the third century B.C., Theophrastus, an Ionian philosopher, wrote of his awareness of climate and vegetation relationships. He noted that vegetation patterns change systematically as one moves up a mountain. He also observed that these same altitudinal "zonations" of vegetation mimic the shifts in patterns of vegetation seen when traveling toward the polar regions of the European continent.[5] Basically, similar climates support similar vegetation. Spruce and fir trees grow near the summits of the Alps and in northernmost Europe (Sweden and Finland). But what if the climate had somehow changed? What would have happened to the vegetation then?

The regularity of the appearance of vegetation in a particular climate regime, regardless of whether it is on a mountaintop or at a high latitude, leads us to expect that a change in climate should cause vegetation to shift its position—up or down on a mountain and north or south on a continent. The vegetation might change its location, but we would expect it to maintain its appearance and species composition.

Thus, our intuition is that the vegetation response to climate change in a mountain region would be for the vegetation zones to move up the mountain if it became warmer, and down the mountain if it turned cooler. For each zone the vegetation should change in altitudinal position but not in fundamental character. Under a different pattern of climate in the past, one might expect a different pattern of vegetation on the landscape, but the vegetation types would be the same as those of today. While the maps of past versus present vegetation might vary, the *legend* of a past vegetation map would be the same as that of a current vegetation map. However, the material in packrat middens indicates that this is not the case; the vegetation in the past map is not comparable to the present map.

Kenneth Cole analyzed the contents of recent and older packrat middens to determine the changes in the distribution of plants at various elevations in Arizona's Grand Canyon over many millennia (Figure 13).[6] Twenty-four thousand years ago the Earth was in full glacial condition. The tops of North America and Eurasia were covered with great continental glaciers that were as much as 2 miles (3 km) thick. So much ice was in these glaciers that the sea stood at least 330 feet (100 m)

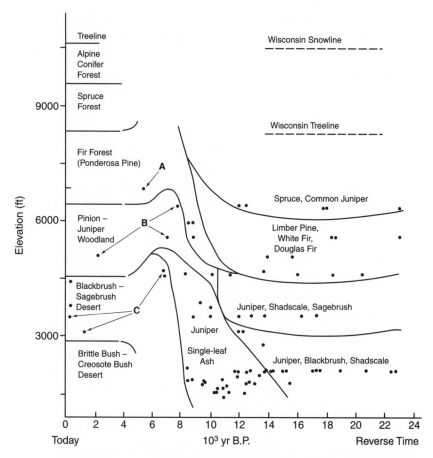

Figure 13. Changes in the vegetation zones of northern Arizona during the past 24,000 years, based on plant and macrofossil data from carbon-dated packrat middens of the Grand Canyon. The dots indicate middens from different altitudes, built at different times. For example, A indicates an upper-elevation midden found at about 7,200 feet with an age of 6,000 years. The pieces of this midden indicate that it was collected in a zone with a nearby fir forest. B indicates middens having contents consistent with pinyon pine and juniper woodland; C, middens with components associated with blackbrush-sagebrush desert. The vegetation zones that are present today were absent in the past and vice versa. From K. Cole, Past rates of change, species richness, and a model of vegetational inertia in the Grand Canyon, Arizona, American Naturalist *125 (1985):289–303.*

shallower than today. The packrat middens inspected by Cole revealed the history of the vegetation from different altitudes in the mountains of Arizona. These glimpses into the past showed the effects of a climatic global warming—the changes when a cold, glaciated world warmed became the world we know today.

How was this vegetation different from what we currently see? First, to get an impression of the modern vegetation, imagine we are hiking up a mountain near the Grand Canyon. At its foot we would start in a shrubby desert dominated by brittle bush (*Encelia farinosa*) and creosote bush (*Larrea divaricata*). As we moved to slightly higher elevations (3,000 feet, or 900 m), the desert would begin to have more sagebrush (*Artemisia* sec. *tridentatae*) mixed with blackbrush (*Coleogyne ramosissima*). A mile up in the mountains (1,500 m), we would stroll through open woodland with pinyon pine (*Pinus edulis*) and juniper (*Juniperus osteosperma*) trees. After trekking through this woodland still farther up to 7,000 feet (2,100 m), we could shelter in the shade of a white fir (*Abies concolor*) or ponderosa pine (*Pinus ponderosa*) forest if we picked a spot where a forest fire had once blazed. Spruce (*Picea engelmannii*) forest would begin at 8,850 feet (2,700 m), and finally, at 10,000 feet (3,100 m) or so, alpine conifer forest with species such as limber pine (*Pinus flexilis*) would take over. At the timberline—slightly below 11,500 feet (3,500 m)—stunted limber pine trees would give way to alpine tundra. Our hike would have brought us through a zonation of vegetation caused by the local climate at different elevations. (If we had picked through any packrat middens along the way, they would have been made of twigs and bits of the nearby vegetation.)

The vegetation implied by inspecting old packrat middens can recreate our path up the mountain in times past. The land would look different. Communities exist today that did not exist in the past, and vice versa. The various plant species that constituted the vegetation in each of the mountain zones changed independently of one another. In the climates of the past, the vegetation did not simply move its position on the landscape. In some cases it rearranged itself into novel patterns that are not at all in evidence today.

Some 6,000 to 8,000 years ago, our hike up the Arizona mountain would not have begun in desert vegetation. Indeed, the blackbrush-sagebrush desert and brittle bush–creosote bush desert at the start of today's hike (and usually found in the lower elevations in this part of Arizona) would have been totally absent. We would have started our imaginary hike through vegetation composed mostly of juniper and single-leaved ash (*Fraxinus anormala*). This combination of species is a *nonextant* type; it does not occur today.

Deeper in time, 10,000 to 12,000 years ago, two other nonextant types of vegetation occupied the lower elevations. These comprised different combinations of juniper, blackbrush, shadscale (*Atriplex confertifolia*), and sagebrush (Figure 14). The pinyon-juniper woodland that today is located between 5,000 and 7,000 feet (1,500–2,100 m) was absent at any elevation. A very different forest with a curious mixture of species, including limber pine, white fir, and Douglas fir (*Pseudotsuga menziesii*) occupied the present-day elevation. A spruce–common juniper (*Juniperus communis*) forest was at the high altitudes where fir forests and ponderosa pine forests are found today.

From ancient plant material hidden in the nests of packrats, vegetation reconstructions challenge the idea that unchanged plant communities in the western United States moved up and down the elevation gradient in response to changes in climate. Other paleoecological analyses using other lines of evidence have demonstrated the same sorts of changes for other regions. What have we learned about the past from these and other studies? "Quite a bit" would be the best answer.

Much of what we have discovered differs from our traditional understanding of constancy in natural ecosystems. The regular patterns of vegetation observed in space by Theophrastus in the third century B.C., and reiterated by philosophers and plant geographers since, are not constant through time. Because understanding the past responses to environmental change is wisdom crucial to predicting future responses of ecosystems to new changes, it is useful to inspect other examples of paleoecological detective work.

How do scientists reconstruct the past? Many kinds of physical evidence are available to them. Small pollen grains from flowering plants fall into lake sediments to preserve a record of the plants around the lake. Microscopic ocean plankton die, sink to the ocean floor, and locked in the deep ocean sediments leave a fossil indication of the ocean conditions at the surface. Beetles fall into lakes and their resistant wing covers remain for millennia to reveal what sorts of insects once flew near the lake. Mammals and birds drop into caves and their bones survive to record their presence. Small rodents such as packrats hide bits of plant matter in nests to leave a history of the vegetation at some particular point in space. Other investigations, including the use of atomic differences in materials to register climate conditions, yield clues to the physical environment of the past. Glaciers accumulate from snowfalls and record the atmospheric gases of long ago, trapped as small bubbles in their ice.[7]

Reconstruction of the past from these bits of evidence is essentially a form of scientific detective work. The attention to detail, logical erudition, assemblage of observations, and eventual remarkable conclusions all seem the work of a modern-day Sherlock Holmes. Much of this work has been used to gain an understanding of the Earth's history during the past 2 million years. This period is referred to geologically as the Quaternary, a particularly active time in the sense of climate changes. Its dynamism has shaped the current distributions and patterns of plants and animals on the terrestrial surface. Within the Quaternary are two subintervals or epochs. The Holocene epoch dates from the present to about 10,000 years ago, and the Pleistocene epoch from 10,000 to about 2 million years ago. Periodic formations of continental-scale glaciers in the Pleistocene gave it the popular name, the Ice Age.

What have we learned from the new tools and techniques for understanding the past at different temporal and spatial scales? One valuable tool uses the determination of subtle differences in the atoms that compose past materials. For example, atoms of oxygen exist as two stable isotopes that are denoted ^{16}O and ^{18}O. The nucleus of the ^{18}O atom has two more neutrons than the eight neutrons of the more common ^{16}O atom. Atoms of these isotopes behave similarly in a chemical sense, but the different masses of the atoms create different effects. For instance, water made of ^{18}O does not get incorporated into ice crystals quite as rapidly as water made with ^{16}O. The ratio of ^{18}O to ^{16}O in the liquid waters of the planet indicates the total amount of ice on Earth.

The ratios of the stable isotopes of oxygen from fossils in deep ocean sediments and from the ice in glaciers in Greenland and Antarctica provide a long and detailed record of the volume of ice on earth (Figure 14A).[8] This record shows that over the past several millions of years, regular cycles have taken place in the inventory of ice on the surface of our planet. Predictable alterations in the Earth's orbit apparently drive these pulsating changes in climate.

In 1941 M. M. Milankovitch mathematically derived expected cycles of change in climate.[9] He theorized that there should be periodic cooling and warming of the Earth's climate driven by small variations in the Earth's rotation and its solar orbit. His predictions included: a 105,000-year cycle in the degree of "roundness" of the elliptical orbit of the Earth around the sun (called the eccentricity of the Earth's orbit); a 41,000-year periodic "wobble" in the spinning axis of the Earth; an additional 21,000-year cycle that determines whether the water-filled Southern Hemisphere or the more terrestrial Northern Hemisphere is closer to the sun in the winter.[10] Milankovitch's cycles cause small variations in the amount of heat coming to the planet and drive the ups and downs of ice formation on Earth over the past several million years. A million years of this history of ice is shown in Figure 14A.

Milankovitch's cosmological variations produce substantial changes to the climate of our planet. The climatic changes through the most recent of the Milankovitch cycles have been documented by scientists who have extracted mile-long ice cores from the central glaciers of Greenland. These cores have annual markers in the glacial ice. Counting back through the Greenland cores, one year at a time, scientists can measure the isotope ratios of snow that fell in a particular year well over 100,000 years ago.

The changes in the amount of ice on Earth over the past 100,000 years give evidence for an initial warm period followed by an extended period of cold temperatures and large volumes of ice lasting from 60,000 years ago until about 12,000 years ago, when the ice appears to have melted and the temperatures increased (Figure 14B). This ice was tied up in great continental glaciers as much as 2 miles (3 km) thick that lay across the top of the Northern Hemisphere.

South of these great glaciers, the majority of Eurasia and North America was covered by an ecosystem in place then but not now—a mixture of tundra sedges and prairie grasses. Because of the odd mixture of plants, this vegetation is called the tundra-steppe. A remarkable and extremely diverse assemblage of giant mammals with equally large predators grazed the tundra-steppe. Nearer the equator, the extensive tropical rain forests of today were shrunk into small areas (called refugia), and savanna-like vegetation was widespread in their place.

Because of the amount of ice stored on the land as glaciers, the oceans were lower. North America was connected to Asia across the

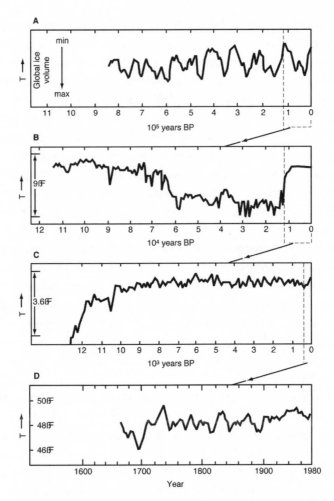

Figure 14. Climate variation. Each panel is a detailed expansion of the panel above it. A, Global ice volume over the past million years (deduced from oxygen isotope variations of planktonic foraminifera in a deep-sea core); B, Global mean temperature variations over the past 100,000 years (based on oxygen isotope variations in a long ice core extracted from a glacier in Greenland); C, Global mean temperature variations over the past 10,000 years (based on oxygen isotope variations in the same Greenland ice core); D, Thermometer readings from England for the past 300 years. From J. B. Harrington, Climatic change: A review of causes, Canadian Journal of Forest Research 17 (1987):1313–39; redrawn from B. Saltzman, Climatic systems analysis, Advanced Geophysics 25 (1983):173–233. *With permission of Elsevier.*

modern Bering Strait. The lowered sea levels caused many of today's islands to be connected to one another. Australia joined New Guinea; the North Island of New Zealand joined the South Island; Japan was part of Asia, as was much of Indonesia. Britain was joined to Europe; the Thames and the Rhine were both tributaries of a river now underseas.

Many present-day islands are parts of the mainland whose lowland connections were drowned when the seas rose again. About 12,000 years ago the glaciers began to melt (Figure 14C); the oceans filled with their melted water. The plant species constituting Earth's vegetation migrated and interacted to form today's vegetation pattern. Huge numbers of large mammals went extinct. Humans armed with new, constantly improving technologies spread across the Earth and instigated a vast alteration of the planet's ecosystems that continues to the present.

As we consider finer and finer time slices through the recent history of the Earth's climate (obtained by interpreting different climate indicators from sea sediments, glaciers, and the historical observations shown in Figure 14), almost all time scales show variation. Regular changes in climate also occur at both the millennial and century markers.

The climate and other Earth systems interact. For example, the location of landmasses near the poles seems to be a requisite condition for continental glaciation.[11] Volcanic eruptions that inject dust and gases into the atmosphere can alter climate for days, even years, depending on the magnitude and height of the injections.[12] The climate can alter the vegetation and ice cover of the Earth's surface, but changes in the Earth's surface can also change the Earth's climate. For example, the amount of ice on the surface of the Earth interacts strongly and reciprocally with climate; the same may be true of feedbacks between the terrestrial surface and the atmosphere.

Because we are today altering the surface of the Earth and shifting the chemical and physical properties of the atmosphere, a current scientific challenge is to understand if our alterations are large enough to affect the surface feedbacks with climate and thus cause a global change in the environment.

Another set of clues for climatic reconstruction comes from pollen grains. The dust-like grains that cover automobiles and cause sneezing in hay fever sufferers every spring are a major source of information about the past. Fertilization in flowering plants is accomplished by transferring the male pollen grains to egg cells in the female part of the flower. The transfer can occur by several means (insects, birds) but in a wide variety of plants the pollen grains are released from stamens to be carried by the wind to the recipient flowers. Those of us who suffer from allergies to pollen are well aware that the production of pollen is seasonal and that some pollens travel farther than others. Wind pollination is relatively inefficient, and a great number of the pollen grains are lost on the way to the recipient flowers. Some of these lost grains fall into lakes, where they sink to the bottom with the lake sediments. These conditions preserve the pollen grains and provide an indication of the past vegetation around the lake.[13]

Pollen analysis, or palynology, has provided considerable insight into past vegetation and the changes in vegetation involved with the climate variations in the Quaternary, particularly the past 10,000 to 20,000 years. One of the findings is that individual species of plants change their ranges independently over time. The result in some cases is novel ecosystems for which there is no modern analogue—a phenomenon that is evident in the record left by the Grand Canyon packrats (Figure 14). The analysis of pollen in sediments and other paleoecological clues have given us a vision of natural systems of the changing world that are strongly unlike those of today. As ecosystems assemble and reassemble over large regions, unique assemblages appear and disappear over time.

The changes in past ecosystems are surprising because until recently many of our ideas about their intrinsic qualities were flavored by concepts rooted in the philosophy and writings of eighteenth-century intellectuals who saw nature as a divine creation of remarkable detail. We considered this view in Chapter 1, with Jefferson's expectation that species did not become extinct and that individuals of the fossil *Megalonyx* species he had found were still alive, perhaps in his new Louisiana Purchase.

God's attention to detail in the Creation was manifested in the precise fitting together of the pieces—the plants and animals that make up the natural systems. Extinction was an affront to the Creator. D. K. Grayson, in a rich historical review of the topic of extinction, quotes Pope's *Essay on Man* as a prime example of this point of view:[14]

Vast chain of being, which from God began,
Natures aethereal, human, angel and man,
Beast, bird, fish, insect what no eye can see,
No glass can reach! From infinite to thee,
from thee to Nothing!—On superior pow'rs
Were we to press, inferior might on ours:
Or in the full creation leave a void,
Where one step broken, the great scale's destroy'd:
From Nature's chain whatever link you strike,
Ten or ten thousandth, breaks the chain, alike.

Such arguments for constancy and design in nature, and for an unbroken natural chain, survive in popular concepts of the "balance of nature" and in a belief in the antiquity and unchanging nature of ecological systems such as the "forest primeval."[15] We see neither constancy nor unbroken chains in the past record.

Even though educated people today have little difficulty believing that species have become extinct or that environs were quite different in a previous time, the concept that a wilderness or constant primal state is "natural" still pervades many of our policies on managing parks and nature preserves. Ecologists know that environments were different in the past. Thus it is surprising that ecological theory abounds with analyses of what ecological systems should be at equilibrium, and that ideas about ecological succession and landscape are posed in terms of a constant environment. Because of the importance of appreciating the omnipresence of change, let us consider a few more examples.

During intervals in the Quaternary when glaciers advanced, no trees existed in England owing to severe climatic conditions and the presence of an ice cap.[16] During the interglacials (when the glaciers were in retreat), tree species migrated from locations south of the Pyrenees and the Alps to the British Isles.[17] Warming climates led to a race among the spreading tree populations to see which could colonize Britain first. The forest composition of England was different during the last four interglacials: species arrived at different times to revegetate the formerly glaciated countryside. Some of the species were there in some of the interglacials, but not in others. Also, the sequence in which tree species become predominant varies among interglacials.

What could have caused the variety of British vegetation? One possibility is differences in the seasonal patterns of temperatures and precipitation that altered the competitive abilities of species.[18] The interglacials could have had variabile temperatures and precipitation even if the yearly averages were the same. Second, chance events involving seed dispersal could mean that a species in one interglacial (but not another) "got lucky" and became established before a competitor species had the chance to do so. Third, there could have been differences in the local timing of the sea-level rise that cuts the British Isles off from the rest of Europe, and in the climate of Western Europe. The English Channel forms when the seas fill from the melting of continental glaciers all over the world, and from the expansion due to warming seawater. Plants migrate across the channel in response to the regional terrestrial climate. During one interglacial, the channel might fill while the adjacent continent was cool; another interglacial might be warmer or drier while the English Channel drowned. Either occurrence would affect which species could arrive to colonize England.

Certainly, the natural vegetation of Britain inspired English pastor-naturalists of the nineteenth century to generate sermons on the detailed design of nature and the wonders of God's Creation. But in fact, these ecosystems had assembled themselves out of the recent geologic past. Other versions of the English vegetation, probably equally marvelous, had occurred in the only slightly more distant geologic past.

The present-day predictability of ecosystem patterns over large areas and the apparent complexity of ecosystem interactions led many ecologists to assume that ecosystem components had evolved together over very long periods. Because species composition is predictable over large areas and the species act as co-evolved units, the discovery that many modern terrestrial ecosystems have assembled themselves in relatively recent time also was surprising to a number of modern ecologists.

Fossil pollen samples from lake sediments have been used with success to develop maps of change in the ranges of eastern North American trees over the past several thousand years.[19] These maps document independence in the ranges of major tree species. As the species migrated across eastern North America following the retreat of the continental glaciers, they moved independently of one another with different velocities and patterns of migration. Comparable maps developed for Europe also show independence of movement for species recolonizing Europe.[20] Accordingly, novel ecosystems with combinations of species not found today covered the landscape at different places and at different times.

As an example, consider today's boreal forests in Eurasia and North America. The strong similarity in the structure and landscape pattern of boreal ecosystems strikes visitors to the far north of both continents.[21] The foreign boreal forest looks familiar. As expected, birches (*Betula*) and aspens (*Populus*) are found in successional stands following fires; mature stands of spruce (*Picea*) and fir (*Abies*) form dark conifer stands; sandy or drier locations support pines (*Pinus*).

There are exceptions: the vast Siberian larch (*Larix sibirica*) forest of Russia is one obvious case. Nevertheless, the broad patterns of boreal landscapes are quite similar. A Russian ecologist can view a Canadian landscape for the first time and confidently "read" the vegetation to see the history of wildfires over the past several hundred years. On inspection, the global boreal forests offer a unity of pattern—a forest ecosystem adapted to fire and cold climate across the top of the Eurasian and North American continents.[22]

However, paleoecological reconstruction of the ranges of the major tree species of the North American boreal forest for the past 18,000 years provides a different view.[23] The relative abundances of the species that make up the forest have changed greatly. Indeed, many of the species combinations that today co-occur and behave so predictably previously were virtually disjoint. Spruces were in one location; pines in another. Alders (*Alnus* spp.) were much more abundant 8,000 years ago. Birch is common today, but not in the past. In these reconstructions one finds unique mixtures of trees and other plants. Analogous cases can be demonstrated for the ecosystems in tropical, temperate, or arctic environs.[24]

Along with striking change in global vegetation, the end of the Pleistocene produced a substantial extinction of many of the large animals in Eurasia, the Americas, Australia, and elsewhere.[25] Because of the size of many of these creatures, they are called the "megafauna." In North America about 10,000 years ago, almost forty genera of large North American mammals became extinct over a short interval.[26] A zoo full of the species of large Pleistocene mammals would need to be more than twice the size of most of today's zoos. Climate change, actions of prehistoric people, and coincidence of the evolutionary turn-overs of species are all candidate causes for this extinction,[27] but they could very well have worked in concert with one another.[28]

Regardless of the cause of these extinctions, accounts of the mega-faunal species stir the imagination.[29] Large carnivores (huge bears, sabre-toothed tigers, cave lions, dire wolves, dog- and hyena-like ani-mals of various sorts) were distributed throughout Europe, Asia, and North America. An array of herbivores (long-necked camels, giant beavers the size of modern bears, horned giraffes, giant armored ar-madillos, huge ground sloths the size of small elephants) grazed vege-tation in North America and Eurasia that currently supports nothing resembling such creatures.

For example, today there are two elephants, the African elephant (*Loxodonta africana*) in Africa and the Asian elephant (*Elephas max-imus*) in Asia.[30] In the late Pleistocene, these two species existed in Africa and Asia respectively, but the African elephant also occurred in Europe along with another elephant, the woolly mammoth (*Mam-muthus primigenius*). North America had three elephants (Figure 15): the woolly mammoth, the imperial mammoth (*M. imperator*), and the Columbian mammoth (*M. columbi*). Asia also had three elephants, in-cluding the woolly mammoth and the modern elephant.[31] Along with these species lived a diversity of related genera from other subfamilies of elephant-like creatures, notably the mastodon (*Mammut america-num*).

Near the end of the Pleistocene, over half the species of large mam-mal herbivores in North America, South America, and Australia be-came extinct. Thirty-seven percent were lost from Eurasia. Africa, where humans evolved in the presence of these large animals, experi-enced only about a 10 percent loss of species.[32]

Megafauna such as the elephant strongly alter the habitats in which they occur. From direct observations in Africa and Asia, the effect of herds of elephant and other large herbivores on the vegetation struc-ture and composition is significant.[33] Elephants in Africa are a major factor in changing the dominant tree species in the vegetation. Areas of relatively high elephant abundance have elevated levels of tree mortal-ity and damage. Grasslands are converted to closed canopy woodlands where elephants are excluded. It is likely that the effect of these crea-tures on the past vegetation in other regions was equally substantial.

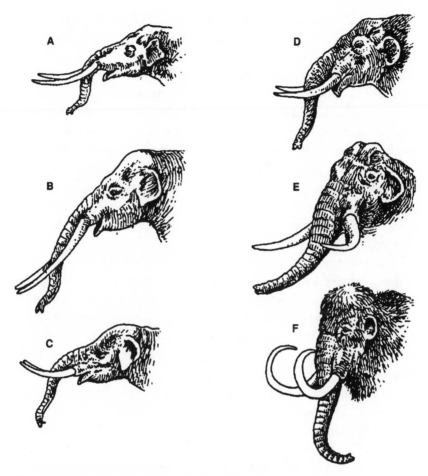

Figure 15. Extinct proboscideans from the late Pleistocene. A, Cuvieronius;
B, Stegomastodon; C, Haplomastodon; D, Mammut; E, Mammuthus
meridionalis; F, M. primigenius. *The last records for the* Cuvieronius,
Stegomastodon, *and* Haplomastodon *are from South America between 11,000
and 13,500 B.P. Mammut (the mastodon) was last recorded in sites from North
America with dates between 9,000 and 12,000 B.P. Similar dates (about 11,000 B.P.)
hold for most recent records of mammoths (Mammuthus spp.) in Eurasia and North
America. Dates from E. Anderson, Who's who in the Pleistocene: A mammalian beastiary*
(pp. 39–81), in P. S. Martin and R. G. Klein, eds., Quaternary Extinctions: A
Prehistoric Revolution, *(Tucson: University of Arizona Press, 1984). Illustrations
redrawn from Anderson.*

The recent disappearance of so many large mammals probably has altered the Earth's vegetation and the kinds of plants that we would expect to be successful.

The so-called Late Quaternary Extinction of large mammals was worldwide but occurred at slightly different times depending on location.[34] Overall, half of the 167 genera of large animals (animals heavier than 100 pounds or 44 kg) on the Earth's continents became extinct.[35] The loss of the megafauna was quite abrupt in North and South America compared to Europe. The variation in location with time supports a premise that the migration of humans out of Africa has had an important role in the extinction of these creatures. In what has been termed a dreadful syncopation, *Homo sapiens* arrives in a location and large mammals become extinct there (Table 1).[36]

In specific cases there is latitude for debate about the role of humans in the demise of a megafaunal species. Reconstruction of past events from bits of fossil evidence is rarely clear-cut. Implication of humans in the megafaunal extinctions involves relatively precise timing of events in the deep past. Is the accepted date for human presence in an area truly the earliest, or have earlier remains of humans not yet been found? Similarly, the last fossil located is unlikely to be the *very* last individual of the species. How long, then, did a particular species "hang on" after its last fossil was produced?

The processes that cause a fossil to form are not random. How does this fact affect interpretation of the numbers of fossils and the timing of the deposition of fossils? How precisely did these extinctions occur? Were they products of human hunting of the animals? Or were they a result of the animals introduced by humans (dogs, rats), of human technologies (ignition of wildfires), of diseases vectored to a naive fauna by humans or the animals that attended them, or of complex combinations of these factors in conjunction with climate change and other events?

People in hunter-gatherer societies can produce manifest changes to their environment—probably to a greater degree than most of us would imagine. Species extinctions that have occurred on remote islands in fairly recent times (Table 1) provide insight into the ways extinctions may have occurred in the past, as we shall see in Chapter 8.

If stone-age cultures produced large-scale extinctions and other changes, then our current technological-industrial culture appears to

Table 1. The "dreadful syncopation" of human arrival and the disappearance of large animals, the megafauna

Place	Time	Consequences
Continents: First Contact with Humans in Prehistoric Times		
Europe and Africa	Late Quaternary time—past 100,000 years	Human presence from earlier times. No major episodes of extinction of megafauna, but some losses.
Australia and New Guinea	Humans arrive 40,000 to 60,000 years B.P.	Major extinction episode follows human arrival and extends to about 15,000 B.P.
North and South America	Humans arrive 12,500 B.P.	Major rapid extinction episode occurs, terminating in about 10,500 B.P.; relatively little extinction thereafter. 74 genera lost from North and South America.*
Islands: First Contact with Humans in Prehistoric Times		
Mediterranean islands	Humans arrive 10,000 B.P.	Major extinction episode occurs that terminates around 4000 B.P., with few extinctions later.
Caribbean islands islands of the Antilles	Humans arrive 7000 B.P.	Major episode of extinction occurs, extending to A.D. 1600.
Madagascar	Humans arrive 2000 B.P.	Major episode of extinction occurs, terminating in A.D. 1500.
New Zealand	Humans arrive 800 to 1000 B.P.	Major episode of extinction occurs that terminates around A.D. 1500.
Islands: First Contact with Humans in Historical Times		
Mascarenes (Mauritius, Reunion)	Humans arrive A.D. 1600	Major episode of extinction occurs that terminates A.D. 1900. Literally hundreds of analogous cases can be cited.[†]
Galápagos Islands	Humans arrive A.D. 1535	Extinction rates elevated two orders of magnitude over background.*

Source: From data provided in R. D. E. McPhee and C. Flemming, Requiem Æternam: The last five hundred years of mammalian extinction (pp. 333–372), in R. D. E. McPhee, ed., *Extinctions in Near Time* (New York: Kluwer Academic/Plenum Publishers, 1999); additions and comments indicated in notes.

*P. S. Martin and D. W. Steadman, Prehistoric extinctions on islands and continents (pp. 17–55), in McPhee, *Extinctions in Near Time.*

[†]D. W. Steadman, Prehistoric extinctions of Pacific island birds: Biodiversity meets zooarcheology, *Science* 267 (1995):1123–31.

be altering the performance of global-scale chemical and physical systems. Because of the changes we have made in the planet's atmosphere, we seem to be heading toward a global environment for which we have little or no historical precedent.

A striking change associated with modern human society is the higher level of atmospheric carbon dioxide associated with the increased burning of fossil fuels (coal, petroleum, natural gas) since the Industrial Revolution.[37] Carbon dioxide is an essential component of plant photosynthesis. Therefore, increases in carbon dioxide levels immediately raise the question of what these changes do to plants and their functioning.

Stomata are the microscopic pores on leaves that open to allow carbon dioxide to diffuse into the leaves for photosynthesis. Inspecting the leaves from historical plant specimens in the Cambridge University Herbarium, F. Ian Woodward found about 40 percent more stomata on the underside of leaves of several species collected 200 years ago, in a time prior to the Industrial Revolution.[38] As the carbon dioxide in the atmosphere increased over those 200 years, the number of stomata in Woodward's sampling decreased.

Plants growing in a time when there was less carbon dioxide in the air had more stomata on their leaves. Having more stomata potentially means that more carbon dioxide can diffuse into the leaves. Thus, more stomata are an appropriate adaptation to lower carbon dioxide levels. To test the cause of the change in his 200-year-old leaves, Woodward duplicated his observations by growing plants in his laboratory in closed chambers with atmospheres containing reduced carbon dioxide. The results indicated that carbon dioxide differences in the air could indeed cause the effect he saw.

The leaves of the past looked different under the microscope; did they function differently as well? The answer involved more scientific detective work. Like oxygen, carbon occurs in two stable forms of slightly different mass. The two stable isotopes are ^{13}C, with one more neutron in its nucleus, and the much more common ^{12}C isotope. The ^{13}C version of the carbon dioxide molecule moves through the stomata of leaves more slowly than the ^{12}C isotope.

The critical clue is that relatively more ^{12}C than ^{13}C shows up in plant tissue when the leaves are more "open" to the diffusion of carbon dioxide. Under these same open conditions, leaves tend to lose more of their internal water. The ratio of ^{13}C to ^{12}C in plant tissue gauges the amount of carbon gained relative to water lost from the plant (plant water-use efficiency). It is a valuable index of plant performance. For example, farmers labor to get as much crop as possible (carbon-in) for each unit of irrigation water (water-out).

When Woodward determined the ratio of ^{13}C to ^{12}C preserved in herbarium leaves inspected in his stomata counting, he found that the water-use efficiency of plants increased after the Industrial Revolution.[39] Why is this significant? It implies that, with respect to the crucial processes of photosynthesis and water use, plants may be functioning differently today than they did 200 years ago: they should be using water more efficiently.

Sciences have "grand challenges" that command and focus a research field: For physicists, what is the nature of the atom? For biologists, what is a gene and how does it work? For astronomers, how did the universe originate? Developing a capability to predict the dynamics of ecosystems in light of our knowledge of the past is another such scientific question, one that could steer ecological research for generations. The evidence from the past challenges our understanding of the natural world. Past ecosystems could have been different in composition, in appearance of the plants, even in the basic manner that the plants functioned. Ecosystem prediction is a grand challenge for ecologists.

Alteration of the planetary environment by today's technological society adds urgency to the challenge of predicting ecosystem dynamics. Recent analyses of the Earth's temperature over the past thousand years (Figure 16) indicate that the average temperature has in the past decade become sufficiently warm to be outside the envelope of variation seen over the past millennium.[40] The cause is not known, but the pattern and its magnitude are consistent with a climatic change in response to human alterations in the chemical nature of the Earth's atmosphere.

The Earth receives light from the sun and radiates heat back out to space as longer-wave-length radiation. If the Earth had no atmosphere, the surface would attain a temperature at which the heat radiated to space would balance that received from the sun. It has been known for quite some time that this phenomenon would occur when the Earth's surface reached a temperature of 1.4 degrees Fahrenheit ($-17°C$). Be-

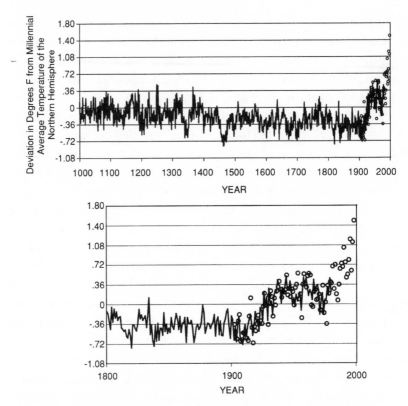

Figure 16. Above, *Reconstructed temperature of the Northern Hemisphere over the past thousand years. The solid lines are the yearly average temperature based on a variety of paleoecological sources. The small circles are annual average temperatures since 1902 based on meteorological station records.* Below, *The same information shown in greater detail for the past 200 years. From M. E. Mann, R. S. Bradley, and M. K. Hughes,* Northern Hemisphere temperatures during the past millennium: Inferences, uncertainties, and limitations, Geophysical Research Letters *26 (1999):759–762.*

cause of our atmosphere, the actual average temperature of the Earth's surface, 61 degrees Fahrenheit (16°C), is considerably warmer than it would otherwise be.[41] Without the atmosphere, the oceans would freeze.

An increased level of carbon dioxide in the atmosphere alters the heat balance of the earth, even though it is a relatively rare constituent. In the atmosphere, carbon dioxide allows light to enter and it becomes heated by the exiting long-wave radiation. Carbon dioxide is a "greenhouse gas." The atmosphere's heat balance is complex. Clouds absorb long-wave radiation (a greenhouse function) but they also reflect incoming radiation back to space (a cooling function). Other atmospheric components affect the Earth's energy balance as well. Principal among these are methane, ozone, nitrogen oxides, man-made chlorofluorocarbon compounds, and water vapor, all of which act as greenhouse gases.

Complex computer models (called general circulation models, or GCMs) have been used to attempt to understand the global climate and to assess the effects of changes in the atmosphere such as might occur from the observed change in atmospheric carbon dioxide.[42] The GCMs developed by different research groups vary in the details of how important Earth processes are depicted in a given formulation. Different assumptions change the approximations for conditions that cannot be simulated directly in the models (cloud formation, the exchange of energy by the atmosphere and the oceans, or the way ice is formed at sea). The models differ considerably in the details of their predictions, even at the coarse scales to which they are currently developed. While they are different one from another, their internal workings are credible approximations of what we know about Earth system processes. The models converge in that all predict an increase in the Earth's temperature as a consequence of increased atmospheric carbon dioxide. Research predicts a warming in average Earth temperature of about the same degree as the warming since the melting of the glaciers 12,000 years ago. Given the response to change seen in the natural systems over this interval, such a warming—if it occurs—will have fearsome consequences.

The potential climate change is likely to occur in concert with major changes to the Earth's surface. The conversion of land from native vegetation to agricultural and urban uses has greatly altered the landscapes of Europe and parts of North America, Australia, and Asia. Land conversion is continuing to affect the landscapes of developing nations in the tropics and subtropics of Asia, Africa, and Central and South America. There is also forest harvesting in Canadian boreal forests and regrowth in Russian forests. As the patterns of natural landscapes are changed, habitats for plants and animals are altered; certain species are eliminated while others are favored.

There is a great deal of appropriate concern about what these changes mean for the biotic diversity of our planet. Land conversion releases carbon dioxide into the atmosphere. Further, changes in land cover alter the surface properties of the Earth in ways that can also alter local and global climates.

The Earth's environment is dynamic over almost any time scale we might consider, and change is an omnipresent feature of natural ecosystems. Our understanding of ecological systems must include dynamic change as an essential attribute of nature. Thus, conservation must involve more than building fences around nature reserves and parks.

Imagine the difficulties of a Pleistocene park ranger in charge of the Grand Canyon 12,000 years ago. According to the history hidden in packrat nests, our ranger would have seen the vegetation reforming in unique ways; mammals of striking grandeur would be going extinct; whole ecosystems long familiar would be disappearing. Our future may be very different from anything in the past, heightening the need to understand what processes cause the patterns we see in nature. We do not really know what packrats will be sticking in their nests a hundred years from now. But such knowledge is a necessity, if we are to manage our Earth wisely.

5 *The Earthquake Bird and the Possum*

The Earthquake Bird

At first the Mississippi seemed to recede from its banks, and its waters gathering up like a mountain, leaving for the moment many boats, which were here on their way to New Orleans, on bare sand, in which time the poor sailors made their escape from them. It then rising fifteen to twenty feet perpendicularly, and expanding, as it were, at the same moment, the banks were overflowed with the retrograde current, rapid as a torrent—the boats which before had been left on the sand were now torn from their moorings, and suddenly driven up a little creek, at the mouth of which they laid, to the distance in some instances, of nearly a quarter of a mile. The river falling immediately, as rapid as it had risen, receded in its banks again with such violence, that it took with it whole groves of young cotton-wood trees, which edged its borders. They were broken off with such regularity, in some instances, that persons who had not witnessed the fact, would be difficultly persuaded, that it has not been the work of art. A great many fish were left on the banks, being unable to keep pace with the water. The river was literally covered with the wrecks of boats, and 'tis said that one was wrecked in which there was a lady and six children, all of whom were lost.—Eliza Bryan, in a letter from New Madrid, Missouri Territory, March 22, 1816

Eliza Bryan relates some of the immediate effects of the largest of a horrific series of earthquakes that struck in the vicinity of the town of New Madrid on the Mississippi River in what is now called the boot heel of Missouri. Of the several New Madrid earthquakes of 1811 and 1812, the one that Bryan describes in her letter is thought to be the largest in recorded history to strike the contiguous United States.[1] It and two of the earlier New Madrid earthquakes register in the top ten such earthquakes.[2]

In all, as many as five earthquakes occurred over the winter of 1811–1812 with magnitudes near Richter scale 8 in the then-remote region. Earthquakes of this scale are capable of causing severe damage over areas 65 miles (roughly 100 km) or more across. Such was the case for the New Madrid earthquakes. People slept outside that winter for fear their houses would fall on them during a nighttime tremor. The land dropped over most of the area, as much 16 feet (5 m) in some locations. Vast tracts of Mississippi floodplain forest were leveled and several counties near the confluence of the Saint Francis and Mississippi rivers in northeastern Arkansas flooded with the drop in elevation.

In the spring of 1886 Charles S. Galbraith, a collector of birds for the decoration of women's hats, shot a Bachman's warbler (*Vermivora bachmanii*) (Figure 17) near Lake Pontchartrain, Louisiana. It was the first recording of the bird since Dr. John Bachman, a Lutheran minister and naturalist, had collected the first two more than fifty years earlier near Charleston, South Carolina.[3] Bachman had given those two birds to John James Audubon, who described the species and named it after his "amiable friend."[4] Galbraith collected six more specimens in 1887 and thirty-one in 1888.[5]

Subsequently, O. Widmann found the first nest of the species up the Mississippi River from New Orleans, in the "sunken lands" of Arkansas. The New Madrid earthquakes of 1811–1812 created these sunken lands, which became drowned forests and swamps filled with felled trees and tangles. The Bachman's warbler was reported as a relatively common bird when it was found there in 1887.[6]

In 1886, the largest earthquake ever reported in South Carolina rocked the city of Charleston. Considerable associated damage and downed timber ensued in the adjacent swampland. In 1901, Arthur T. Wayne collected a singing male Bachman's warbler near Mount Pleasant, just across the Cooper River from Charleston. It was the first

Figure 17. Bachman's warbler (Vermivora bachmanii). *From J. J. Audubon,* The Birds of America, *vol. 2. (Philadelphia: J. B. Chevalier, 1840).*

record of the species in South Carolina since Dr. Bachman had collected the original two birds in 1886.[7] Wayne's description of the breeding habitat of the bird provides a clue as to why the Bachman's warbler might be associated with the aftermath of earthquakes: "Such places are simply impenetrable on account of the dense blackberry vines, matted with grape vines, fallen logs piled on one another, and a dense growth of low bushes."[8]

The Bachman's warbler has been described as the rarest North American songbird. It is (or was) one of the smallest warblers, having a total length less than 5 inches (10–12 cm). The male is olive green above, with its face and underparts yellow. It has a black throat patch and a black crown patch. The female lacks the black throat; the upper parts are olive green, the forehead and underparts yellow, and the crown grayish. The species is migratory. It arrives early (March) at its breeding grounds in the southeastern United States and departs early (August and September). Most of the specimens have been collected in Florida during transit. Its migratory path funnels it into the Florida Keys and eventually leads to wintering areas in Cuba and the Isle of Pines.

The species nests in low, wet, forested areas where there is an open forest canopy above and extremely thick ground vegetation below. Reported nests in several states conform to the tangled thick vegetation of Wayne's description and frequently involve "cane," a native bamboo (*Arundinaria giagantea*). The Bachman's warbler seems a bamboo or cane specialist in its habitat preference.[9] The demise of cane because of flood control and cultivation, after its historical prominence on the floodplain landscapes, may be a significant element in the disappearance of the bird. Its nests were close to the ground (typically less than 3 feet, or 1 m, above), but the species tended to feed and sing from high in the canopy. The warbler's habitat seems to be a part of the gap-replacement cycle that occurs in the destruction and early recovery after the fall of a large tree (or of several trees in proximity). In addition, a relatively specific environment situation is required: wet, with thick, tangled ground vegetation or cane.

The Bachman's warbler had a surprising ability to appear periodically in large numbers. In addition to Galbraith's shooting of thirty-one of the birds in one year near New Orleans, J. W. Atkins captured fifty-eight specimens in Key West during the migration period in July–August of 1888 and 1889, and apparently saw as many as two hundred.[10] W. Brewster reported collecting forty-six birds and seeing large numbers in migration (March 1890) in Florida, while Wayne collected fifty over two years in Florida (1892–1893).[11]

Audubon Field Notes (now called *North American Birds*) is a journal that publishes the more unusual bird observations made by thousands of field ornithologists and birders in a given season. By the middle of the 1950s, *Notes* was reporting at most one or two observations of Bachman's warblers annually (in sharp contrast to the hundreds seen and collected by single observers a few decades earlier). The last recorded observation in the United States was in 1988 in Louisiana. Eight reports of the species came from its Cuban wintering grounds between 1978 and 1988. Today the species is quite possibly extinct; at the very best, its population levels are extremely low.

By current standards, the collections of Bachman's warblers by some well-known early American ornithologists were rather large.[12] Furthermore, several aspects of the bird's natural history worked against it. Its nests, close to the ground, were vulnerable to predators. Its migration route passed through a hurricane-prone area where an unfortunately timed storm could decimate the migrating population. Also, it migrated to its breeding grounds early enough to be killed by unseasonably cold weather or snowstorms. It was found in locations wet enough to have standing water but not so wet as to be a lake or marsh—a minor subset of most landscapes. Its immediate nesting requirements were perhaps best met by a large canopy gap in the wetter part of a floodplain forest.[13]

In that the Bachman's warbler was such a rare species, the exact cause of its demise will probably never be known. Certainly, many of its former breeding habitats, such as the vast "sunken lands" of Arkansas that have been drained for agriculture, are gone. Its wintering grounds in the Cuban lowlands of cane similarly have been largely converted to agriculture, especially to sugar-cane farming.

The decline of the Bachman's warbler began early in the twentieth century. The last recording of a nest of the species was in 1937. Many of the more recent breeding season records have come from the I'On swamp, near the site of Bachman's original discovery in 1833. It was hoped that the amount of habitat for Bachman's warbler might have improved in the swamp after the forest canopy damage of Hurricane Hugo in 1989. But searches to find the bird there have failed.[14]

In the broad outlines of this warbler's story is a tale of large-scale disturbances and a species dependent on short-lived habitat elements of a dynamically changing landscape mosaic. The interaction is not uncommon; it has led to the success of some species and the extirpation of others. Size of landscape and size of disturbance combine to make certain species inherently vulnerable to extinction—especially in altered landscapes.

We recall that the rarity of the ivory-billed woodpecker sprang from the bird's dependence on specialized feeding sites produced infrequently during a forest's gap-replacement cycle. For such a species, as the area of a landscape increases, so should the amount of usable habitat. Sadly for the ivory-billed woodpecker, the more its habitat shrank, the less feeding habitat was available for it to find.

The Bachman's warbler might be expected to have experienced the same sort of interrelationship. If there were more landscape area, then there should be more potential habitat (and vice versa). However, there is an additional complication: the rare Bachman's warbler would sometimes appear in large numbers; the woodpecker never did.

The feeding niche of the ivory-billed woodpecker was produced by centuries of tree birth, growth, senescence, and death. The Bachman's warbler utilized habitat produced by the fall of large trees in the wetter areas of a floodplain forest mosaic. Violent events, such as hurricanes, earthquakes, or felling trees across large regions, could temporarily yield extensive areas of optimal habitat.[15] If the other requirements of the species were met, the warbler's numbers could leap. When the New Madrid earthquakes leveled forests and the sunken land flooded, the potential habitat of the Bachman's warbler should have momentarily increased, making it at least temporarily a common bird.[16]

Violent events that destroy vegetation and reset vegetation dynamic cycles are called *disturbances.* The size of a landscape and the size of the disturbances to the landscape determine much about the types of species that can sustain viable populations. Indeed, the ratio of landscape size to disturbance size can be used to classify landscape dynamics. In turn, this ratio affects the diversity of species and the likelihood of species extinction.

Disturbances are significant, often abrupt, violent changes in the environment of an ecological system. The scales of both time and space define a disturbance. When changes are small or frequent, they are incorporated into the environment of the ecosystem; when they are sufficiently large and infrequent, they are catastrophes. Disturbances lie between these temporal bounds. The rising of the sun each morning is a large environmental change, but it is sufficiently frequent to be incorporated into biological and ecosystem processes. The explosion of a volcano such as Krakatau is abrupt and violent; collision with an asteroid is devastating. These latter events occur so seldom that species are unlikely to have developed adaptations. In most locations, earthquakes are sufficiently infrequent to be on the edge of the time-scale boundary for disturbances.

Typical disturbances include wildfires, floods, droughts, and extreme meteorological conditions (frosts, hurricanes, extreme winds). Disturbances are agents of death and destruction, but they are also part of the natural environment of an ecosystem, to which species are adapted in varying degrees. Environmental disturbances cause substantial rebuilding in ecological systems: recovery features dynamic changes in vegetation structure, ecological interactions, and ecosystem processes.[17]

Disturbances often occur at spatial scales that are quite a bit larger than the sample units used in ecological research.[18] Also, their recurrence tends to be less frequent than the duration of most studies. For these reasons the effects of disturbance are often missing or poorly estimated in calculations of significant ecosystem processes. Disturbances cause spatial heterogeneity in landscapes and push natural systems away from any equilibrium condition.

Landscapes can be relatively large or small with respect to the size of their disturbance regimes. Large landscapes are big enough to average out the effects of disturbances. There disturbances alter a small portion of the area during a given interval and reinitiate the process of ecological succession and recovery at that site. The landscape as a whole is a mosaic of patches, each patch being disturbed by different events. Forests interacting with wildfire are an example: On a large landscape, a given area of mature forest might be destroyed by fire, while over the same interval other areas previously destroyed are recovering to mature forest. If such a landscape were in perfect balance between disturbance and recovery, there would be an overall equilibrium. The mixture of ages of forest would be the same every year.

In reality, the intensity of disturbances varies from year to year and a landscape is never in perfect balance or equilibrium. However, a landscape large enough to average out the annual variations could be relatively close to ideal equilibrium. Such a case could be termed a predictable landscape or a quasi-equilibrium landscape, reflecting that its overall change consists in small variations around the condition expected for an ideal equilibrium landscape.[19]

Foresters involved in sustainable forestry derive a steady yield of forest products from a landscape by using essentially the same idea. For each tree cut, new trees are planted sequentially to replace those harvested. The area cut each year is based on the length of time required to grow trees to harvestable age (rotation time). The harvest plays the role of the disturbance. The mosaic forest landscape described in Chapter 2 also is a predictable landscape. In that case, the scale of the disturbance is the area of the fall of an individual tree. This small disturbance is averaged out over a large forest and produces the mosaic of the forest canopy.

The ratio of disturbance size to landscape size is a convenient way to separate landscapes from one another according to their dynamics. When the disturbances are large (with respect to the area of the ecosystem), the landscape's attributes are much less predictable. The average condition of such a landscape at any one time usually is not the condition of the landscape at another time. In such "effectively nonequilibrium" or "unpredictable" landscapes, large disturbances drive change. The future condition is knowable only when the disturbance history and the pattern of future disturbances are known. These landscapes are difficult to manage, and it is hard for specialized animals to survive on them.

The Possum

Seventy one lives were lost. Sixty nine mills were burned . . . Mills, houses, tramways, machinery were burned to the ground; men, cattle, horses, sheep, were devoured by fires . . . At midday, in many places, it was dark as night . . . The speed of the fires was appalling. They leaped from mountain peak to mountain peak, or far out into the lower country, lighting the forests 6 or 7 miles in advance of the main fires . . . Such was the force of the wind that, in many places, hundreds of trees of great size were blown clear of the earth.—Judge L. E. B. Stretton, in Report of the Royal Commission to Enquire into the Causes and Measures Taken to Prevent the Bushfires of January, 1939, and to Protect Life and Property. Melbourne, 1939.

The Leadbeater's possum (*Gymnobelides leadbeateri*), an endangered Australian marsupial, is a dramatic example of the difficulties of surviving on an unpredictable landscape. The drama springs from the interactions of an interesting Australia marsupial with some relatively specific habitat requirements, old-growth forest, and large Australian wildfires.

Like the Bachman's warbler, the Leadbeater's possum (Figure 18) has always been a rare animal.[20] Described in 1867, it was thought to be extinct by 1909. The species was rediscovered in 1961. It occurs sporadically over a small range of only 40 × 50 miles (60 × 80 km) in the tall wet eucalyptus forests of central Victoria. As an adult, the possum's body is about 6 inches (16 cm) in length and the tail is a half inch (1 cm) longer than the body. It weighs about the same as two hen's eggs, around 5 ounces (140 g). The species feeds mostly on sap and gum from plants. It licks honeydew, a sugary substance secreted by some insects. It also eats insects and other small arthropods such as spiders.

Much of the possum's sustenance comes from the gum produced to heal the wounds it makes on three species of acacia shrubs found in the understory of the eucalyptus forest habitat. The animal bites U-shaped notches in the trunk and branches of these plants to stimulate the flow of sap and gum. The possum revisits these sites at night to feed. The

Figure 18. Leadbeater's possum, Gymnobelides leadbeateri. From D. B. Lindenmayer, Wildlife and Woodchips (Sydney: University of New South Wales Press, 1996). With permission.

colonial species occurs in groups of up to twelve individuals (typically a mated pair, their subadult young, and two or more nonbreeding adult males). There are two breeding seasons per year with a maximum of two offspring per season (the average is 1.6). The sex ratio in the colonies is 3 to 1 in favor of males. The lack of females, along with the low success in producing young, gives the Leadbeater's possum a very low rate of population increase. Possums interweave nests of bark from living trees with ancient hollow trees and stumps to make their dens. Individuals spend about three fourths of their time resting in the colonial den to conserve their energy.

The Leadbeater's possum is found in old-growth forests dominated by mountain ash (*Eucalyptus regnans*). Attaining a height of over 330 feet (100 m), mountain ash (Figure 19) is the world's tallest flowering plant. Its strong wood allows the trees to reach their remarkable heights, with diameters of 6 to 10 feet (2–3 m). The trees are extremely straight with an open and proportionally small crown. They occur in the higher elevations of Australian mountains. Although they require

Figure 19. Old-growth Eucalyptus regnans *forest, prime habitat for the Leadbeater's possum. From D. B. Lindenmayer,* Wildlife and Woodchips *(Sydney: University of New South Wales Press, 1996). With permission.*

cool summers, they can tolerate frosts (up to eighty per year at higher locations). The species has a small range and is found in the Australian states of Victoria and Tasmania (Figure 20). It is a valuable timber tree used for furniture, trim, and general construction. It also is an important tree for the Australian wood pulp industry.

Mountain ash is in the *Eucalyptus* subgenus *Monocalyptus*. Like

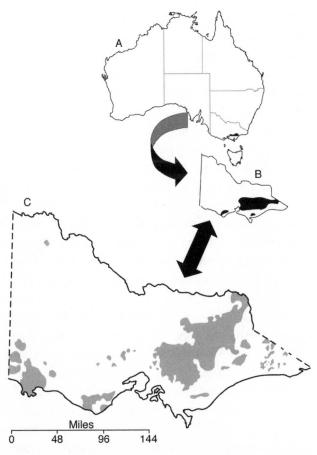

Figure 20. A, *The distribution of mountain ash in Australia. From N. Hall, R. D. Johnston, and G. M. Chippendale,* Forest Trees of Australia *(Canberra: Government Printing Service, 1970). B, Enlarged map of the state of Victoria (ibid.). C, The areas burned in Victoria during the wildfires of 1939. From T. Foster,* Bushfire: History, Prevention and Control *(Sydney: A. H. and A. W. Reed, 1976).*

other monocalpyts (and unlike most other eucalyptus species), it does not sprout when burned or damaged by wildfires. For regeneration, wildfires must be large and so hot that they leave a sterile soil covered with fine white ash. These ash beds, the optimal seeding site, also torch the parent trees. Once the mature trees have been killed, it takes years for seed-producing trees to be found on the site.

While a young mountain ash forest regenerates and grows from the ashes of hot wildfires, it takes twenty to thirty years before any individuals of this generation become large enough to resist light surface fires. During this time the new forest is fire vulnerable. When fires are too frequent, the forests burn a second time before the young trees can produce seed, thereby eliminating mountain ash from the site. Further, if hot wildfire does not burn the site in the lifetime of a tree, then mountain ash is lost from the location. Such an unburned forest is likely to become temperate rain forest.[21]

Wildfires are a vital part of the Australian environment, and Australian plants and animals demonstrate a wide range of adaptations. In wet years, more grassland fires are apt to occur (owing to increased grassland fuels); dry years tend to increase the likelihood of forest fires.[22] In terms of the Leadbeater's possum, it is the forest fires that are of interest. One fire in particular, the "Black Friday" fire of January 13, 1939, illustrates the magnitude of some of the Australian wildfires and is influential in the habitat of the possum even today. The Black Friday fire, described by Judge Stretton, in the quotation that opens this section, was centered mostly in Victoria. A severe drought had taken place in 1938, and difficult-to-control wildfires were already burning in the region.

On January 13, 1939, unusual weather conditions caused the temperature in Melbourne to reach 114 degrees Fahrenheit (45.5 degrees C)! The relative humidity was around 8 percent and the winds were 20 miles per hour (30 km/hr) with gusts of 40 miles an hour (60 km/hr) and higher. Wildfires can have firestorm effects when they are burning dry plant materials under these conditions of temperature, moisture, and wind. Firestorms generate intense heat that flash-ignites trees well in front of the towering flames. Winds roar at speeds of over 60 miles per hour (100 km/hr) into the fire to replace the air being conveyed aloft by heating in the fire's interior. Fire whirlwinds and larger fire tornadoes swirl in the fires and hurl flaming bark and branches as far as 6 miles (10 km), where they ignite other wildfires.[23]

Throughout Victoria on that Friday the thirteenth, existing wildfires were fanned to become a holocaust. In the Black Friday fire and the immediately preceding fires, 5,400 square miles (1.4 million ha) of forest were destroyed (Figure 20). Because of the smoke, ships at sea on the Sydney-to-Melbourne run either anchored or used foghorns as they traveled. Smoke darkened the sky in New Zealand two days later. One victim at a timber mill climbed into an elevated water tank, hoping to escape the fire. The tank was later found to have boiled dry.[24]

Much of the forested land in Victoria was destroyed by the fire of 1939, but mountain ash regenerated well afterward. Thus, 1939 is a landmark in the regeneration of mountain ash, in that much of the Victorian mountain ash forest dates from that year.

It is easy to imagine that the Black Friday fire had an immediate and devastating effect on any Leadbeater's possums in its way. Fortunately, the species (which at that time was thought to be extinct) was not completely wiped out. The historic "Black Thursday" fire of February 6, 1851, which appears to have been of similar magnitude to the Black Friday fire some eighty-eight years later, burned as much as 25 percent of the mountain ash forests and could readily have eliminated the possum as well.

The terrible fires of 1939 eventually generated prime Leadbeater's possum habitat. The large mountain ash trees destroyed in the fire became stumps and snags for possum nests. However, those tree stumps and snags are now falling over. The present-day nesting sites of the possum are in large trees at least 190 years old.[25] The Leadbeater's possum will be severely limited by nest site availability in the coming decades. At the current rate at which the nest trees (mostly trees killed by the Black Friday fire) are falling over, very few will remain by the year 2065. The existing forests regrowing from the 1939 fires will not be old enough to produce suitable candidate nest trees until 2140. The possum seems to be a species that is rather difficult to move from place to place, confounding efforts to preserve the species by shifting it to suitable habitat outside its small geographic range.[26]

Current forestry practice for mountain ash calls for rotations that are not long enough to generate trees of sufficient age to be used by the possum as nest trees. Further, the old-growth mountain ash forests are remarkable natural treasures, and prevention of fires such as those in 1939 will undoubtedly be a continuing land-management goal in Victoria. Management of the interaction of forest dynamics and wildfire disturbances in such a way as to sustain the Leadbeater's possum is a challenge of the highest order. Knowledge of the species' habits, its habitat needs, and increased insight into possible management options are critical. The possum lives on a landscape that is not large enough to average out the effects of the large disturbances that change it—an unpredictable landscape.[27]

Natural landscapes can be either predictable or unpredictable. We have seen that this classification is related to the ratio of the size of a landscape to the size of its disturbances.[28] If the fall of the tree is the disturbance of interest (a gap-scale disturbance), then the watersheds of the smallest streams in the Appalachian Mountains of Virginia should be predictable landscapes, because they are larger by quite a bit than the gap-sized disturbances. However, if Appalachian wildfires are the disturbance shaping the landscape cover, these watersheds are probably too small to average out the effects of wildfire disturbance. As unpredictable landscapes with respect to wildfire, their land-cover condition can only be understood if their fire history is known. Only in the largest parks in Appalachia are the landscapes possibly large enough to compensate for the effects of disturbance arising from typical-sized forest fires. Similarly, forest fires in Russia are large enough to make entire Siberian regions unpredictable landscapes, but even the vastness of Siberia may not be large enough to average out these variations and function as a predictable landscape.[29]

Unpredictable landscapes occur naturally. Siberia may be a case where the entire ecosystem may inhabit unpredictable landscapes. The hurricanes that disturb West Indian forests are large when compared to the size of the islands in the Caribbean. Thus, these islands are small with respect to the spatial scale of their major disturbance and may function as unpredictable landscapes. A similar example would be the spatial extent of floodplain forests compared to the spatial extent of their disturbance, floods.

The ratio of disturbance to landscape characterizes the nature of the dynamic change of the landscape as well as the relative predictability of this behavior. Thus, when humans convert land to agriculture or other use, thereby shrinking the natural cover of the area, the remaining landscape becomes more unpredictable. Alternatively, when humans change the scale of disturbances so that they are larger, unpredictability of landscapes again ensues. Large-scale environmental change, human land-use changes, and natural or human-induced changes in the climate all can alter the size (and timing) of disturbances, and consequently alter the degree to which landscape behavior can be predicted.[30]

Understanding the dynamic response of natural and human-modified landscapes is essential to effective management. When we reduce landscape extent, we erode our ability to predict and manage landscapes. The difficulty in managing Australian landscapes to preserve the Leadbeater's possum is a case in point.

Disturbances are complex in their history, in what causes them to occur, and in how frequently they strike a given location. For example, wildfire is not a purely random event. On Black Friday the fire's intensity was strongly related to the previous environmental conditions, in that the amount of burnable material affects a fire's properties. A second fire following immediately after the one on Black Friday would have found little fuel and would have been a different fire, even if everything else had somehow been the same. These considerations complicate the understanding of landscape pattern as a consequence of disturbance. What do we expect from different disturbances?

First, consider a simple null model of disturbance and land area. If the same amount of disturbance alters the landscape each year, the landscape will have a higher percentage of recently disturbed or "young" areas and much less "old" area (Figure 21). Why? Because some of the land is redisturbed and returned to the young category before it gets very old. The vegetation on a given area must run the gauntlet of each year's disturbances to become old. The larger the percentage disturbed each year, the more young area and the less old area is on the landscape. The more of the landscape disturbed each year, the lower the average "age" of the plots on the landscape.

In a predictable landscape with constant disturbance damage each year, there should be fewer old landscape types than young. Thus, a species that needs old-growth conditions will have less area available to it than a species that can use younger, more recently disturbed habitats. Many of today's rare and endangered species fall in this category. Species requiring habitats that take a long time to develop are intrinsically vulnerable when natural land is diminished or disturbances are increased. The beautiful large-eyed animals that are poster species for

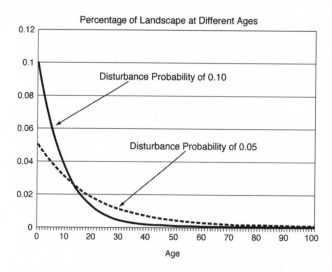

Figure 21. The connection between the probability of disturbance on the ages of the patches that make up a landscape.

conservation groups are often nocturnal creatures that inhabit holes in large trees, habitat that is slow to be created. Protection of owls, lemurs, and possums forces preservation of large tracts of natural landscapes.

If the likelihood of disturbance varies from year to year, the availability of old-growth habitat that eventually develops from the ensuing regeneration will also vary over time. During periods when there is little or no old-growth habitat, dependent species suffer population collapses. This seems to be the current situation with the Leadbeater's possum. The Black Friday fires of 1939 provided a surfeit of den sites, but they will not be replaced for years to come. This difficulty may have been compounded by the terrible wildfires in the Australian Alps during 2002.

For Leadbeater's possum, the scale of the disturbance is large and the scale of the landscapes is small. The habitat of the possum is generated by wildfires and the passage of time. A hot wildfire has the immediate effect of destroying possum habitat at the location of the fire, along with likely destruction of the local possum population. If one could census the entire population of the Leadbeater's possum over the years, one would expect the numbers to feature abrupt drops associated with fires, then a hiatus followed by an eventual recovery as suitable habitat slowly developed over time. Interestingly, this pattern is not unlike the age structure of the continental population of the possum's habitat tree: the age distribution of mountain ash too is uneven, with more trees than one would expect from the 1939 fire regeneration and fewer trees than expected that are older than the destroying fire.

Large disturbances create coarse-grained landscapes made of large patches of habitat of the same age (Figure 22). Small disturbances produce fine-grained landscapes. Some animals and plants are better than others at dispersing across large zones of unsuitable habitat. Efficient dispersers fare better on coarse-grained habitats. Some species require elements from several different habitats. For example, a deer might graze in a young recovering burned forest, but shelter from the cold in older patches of forest with a closed canopy. Fine-grained landscapes would put the diverse habitat elements together in a way that could favor such species. Certainly, the white-tailed deer (*Odocoileus virginianus*) has prospered in the fine-grained habitat created by the predator-free, sprawling suburbs in the eastern United States.

What if a conservation biologist in 1930 had had the task of predicting the viability and location in 1940 of both the mountain ash and the Leadbeater's possum? Even with a perfect knowledge of the species,

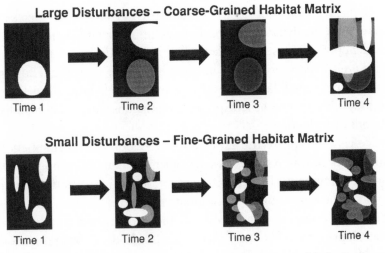

Figure 22. *The size of the disturbance determines the "graininess" of the landscape and the size of the patches.*

without some way to foretell the Black Friday fire of 1939, the state of the population in 1940 was simply not predictable from information of the 1930s. The difficulty lies in the intrinsic unpredictability of land-scapes that are small with respect to the size of their disturbances.

In order to manage rationally, managers of natural landscapes must be able to predict the future response of the landscape to a specific regime of disturbances.[31] The intrinsic unpredictability of small land-scapes vexes the best-laid management plans. In large predictable landscapes, the dynamics of animal populations derive from knowing the birth and death rates of the population. From these rates, a wildlife manager has a good idea of the future of the populations and can de-sign appropriate management strategies. On an unpredictable land-scape, species such as mountain ash and the Leadbeater's possum have unstable population structures. The state, age distribution, and spatial arrangement of the tree and possum are consequences of the recent fire history. Prediction is extremely difficult; one fire can reverse the man-agement plan of decades. Another complication is the population growth rate—the response of a plant or animal population to dis-turbance is dependent on the size of the population growth rate com-pared to the disturbance frequency. When the return interval of dis-turbance is similar to (or shorter than) the time required for a popula-tion to recover, the potential for population oscillations, instabilities, and extinctions all greatly increase.[32]

As the availability of habitat suitable for a given species waxes and wanes with the dynamic responses of an unpredictable landscape, the resident populations are exposed to habitat "bottlenecks": the popula-tions are forced to low levels in times of habitat shortage. Thus, one might expect small landscapes driven by relatively large disturbances periodically to present their species' populations with rather high like-lihoods of extinction.

We can investigate the expected patterns of species numbers for landscapes of different sizes by using simple computer models. One sample model runs a ledger of the vegetation changes over time for ran-domly occurring wildfires in mountain eucalyptus forests. The forest-fire interactions are comparable to those described for the Lead-beater's possum in Victoria.[33] The simulated "data" from this model for different-sized landscapes provide the expected change in habitat for animals requiring different types and ages of vegetation. The larger the landscape simulated, the more species the landscape should contain. This model exercise is a simple, theoretical application of the niche concept (Chapter 3) to disturbance-driven habitat change.[34]

In this experiment, big landscapes have more species (Figure 23). The increase in diversity with landscape size slows as the landscape area becomes larger. This happens with or without modeled competition among the animals using the habitats. Competition does not change the *shape* of the curves; however, there are fewer species on a landscape of given size when the effects of competition are included in the simulation. Small landscapes are extinction machines. They support smaller populations that are prone to extinction due to fluctuations in suitable habitat, which are a product of disturbances and the rates of recovery.[35]

Small landscapes with high extinction rates, and consequently a low richness of animal species, are what we have called unpredictable landscapes. The species that are being lost from these unpredictable landscapes are those that are relatively specialized in their habitat use. Species that require successively older types of habitats are extinction prone. Their preferred mature habitats are rare because frequent

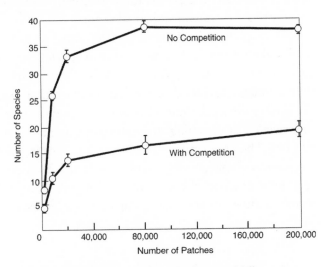

Figure 23. The number of species on landscapes of different sizes after 300 simulated years, with and without interspecific competition. From S. W. Seagle and H. H. Shugart, Landscape dynamics and the species-area curve, Journal of Biogeography 12 (1985):499–508.

disturbance affects recovering landscape patches before mature habitat can develop. Species with relatively low reproductive rates cannot build their populations quickly enough when good fortune provides a surfeit of suitable habitat; these tend to be more prone to extinction. Species whose maximum densities in favorable habitats are relatively low do not have the numbers to survive on a small changing landscape.

Model simulations are abstract representations of a reality that is likely to be far more complicated. Nevertheless, the attributes of the species that become extinct in these simulations are the attributes of the real species that are becoming extinct on real landscapes.

Ecological disturbances create landscape mosaics. This easily observed pattern and its causal processes imply an abandonment of the "balance of nature" view of ecosystems—a view that still dominates discussion, especially public discussion, of conservation and biodiversity protection.

In general, one should not expect nature to be in balance. Striving to achieve balance may not be an attainable or even a logical goal of ecosystem management. Populations vary naturally, but the chances that they will become extinct increase greatly on small landscapes, especially those driven by large disturbances.

Species such as Bachman's warbler and Leadbeater's possum are small stakeholders in the great casino of species survival. Occasionally a windfall comes their way (think of the New Madrid earthquake and the Bachman's warbler) but the chances of winning over time are small. We further diminish their odds by reducing the area of the ecosystems they occupy or changing the disturbance regimes that generate their habitats. The same unpredictability that ensues from shrinking a landscape also complicates any management of such species. Management requires prediction. We lose our capability for prediction and effective management as we shrink and fragment our landscapes.

6 *The Most Common Bird on Earth*

The most abundant species of wild bird is the red-billed quelea (*Quelea quelea*) of the drier parts of Africa south of the Sahara with a population estimated at 10 billion of which a tenth are destroyed each year by pest control units.—*Guinness Book of World Records,* 1988

The red-billed quelea (*Quelea quelea*), an African weaver finch (family Ploceidae) is a small, mostly brown bird (Figure 24) found in the annual grasslands of Africa. Other than ornithologists and the people who see the bird in its native habitat, almost no one has ever heard of the little quelea, although it may be the most common bird on earth. Its annual grassland habitat has an extremely pronounced wet-and-dry seasonality. When the wet season and its rains come, annual grasses sprout from seeds, grow, and mature to produce the seeds for the next wet season. Queleas are adapted to reproduce quickly when the wet-season rains cause fresh vegetation to grow at a particular location. When the dry season comes, quelea moves. The species is highly nomadic; it responds to food shortages by changing location when conditions are unfavorable.

Given the seemingly stressful conditions in the environments where the birds are found, it is surprising that the red-billed quelea survives at all, much less that it is so remarkably successful. Queleas feed on native annual grasses, often of the genera *Echinochloa, Panicum,* and *Sorghum.* Any animal that feeds primarily on the seeds of annual grasses

Figure 24. The red-billed quelea (Quelea quelea). *Courtesy of R. L. Smith, Jr.*

has a potentially major problem—its food supply periodically appears and then disappears. A wide range of adaptations allows annual-seed eaters to survive. For example, rodents in annual grasslands often store seeds in caches to survive lean times, an adaptation that was taken up by human populations once they domesticated and became dependent on annual grasses such as barley and wheat.

If they are to survive, queleas also must cope with the challenging feast-or-famine situation attending annual-grass feeders. One might expect such species to be regularly wiped out (at least locally) when small fluctuations in the growing season result in seed failures of their food. The red-billed quelea copes well enough to be the world's most common bird.

Part of quelea's success arises from its life-history adjustments to its environment. Over the African dry season, the species has to subsist on the ever-diminishing supply of seeds produced by the annual grasses at the end of the last wet season. As food becomes scarcer, the species feeds actively and gains sufficient weight reserves to be able to migrate to more auspicious areas (*A* in Figure 25).

With the onset of rains in the African wet season, the seeds that are eaten by quelea germinate. The resulting severe food shortage lasts six to eight weeks. The birds are forced to migrate to better situations (if they can find them). For red-billed queleas, the savannas in which they

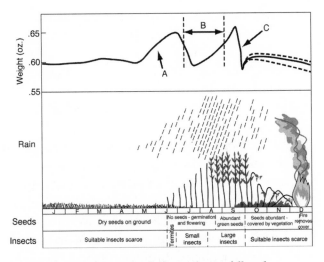

Figure 25. Seasonal changes in body weight, rainfall, and availability of grass seed and insects or food for the red-billed quelea (Quelea quelea) in an area with one wet season per year (Lake Chad, for example). The initial weight increase (A) precedes the "early-rains migration," when quelea typically leave a given locality. The interval between the dashed lines (B) denotes the time the birds are absent from an area. Their weight drops (C) as birds lay a clutch of eggs and start incubating. There is then a weight gain, which varies with the availability of food supplies, the clutch size, and other energy drains on the individuals rearing young. From J. A. Wiens and R. F. Johnston,. Adaptive correlates of granivory in birds (pp. 301–340), in J. Pinowski and S. C. Kendeigh, eds., Grainivorous Birds in Ecosystems (Cambridge: Cambridge University Press, 1977); modified from P. Ward, The breeding biology of the black-faced dioch, Quelea quelea, in Nigeria, Ibis 107 (1965):326–349; idem, The migration patterns of Quelea quelea in Africa, Ibis 113 (1971):275–297; and personal communications.

live are an ever-shifting mosaic of patches of varying suitability according to their recent rainfall history: some areas have dry seeds; others
have immature grasses, mature grasses with abundant green seeds, or
senescent grasses with no seed. By moving over distances of 30 to 120
miles (50–200 km), the birds subsist in an inhospitable universe by
moving from one patch of habitat with suitable food to another.[1]

The so-called early-rains migration (*B* in Figure 25) ends when queleas return to formerly abandoned locations. By this time the rains have
come and grass seeds have germinated. The new grasses mature, and
eventually fresh green seeds become available, ending the local food
shortage. Then queleas establish communal breeding colonies in appropriate locations (*C* in Figure 25). They typically nest in thorny acacia trees. Their breeding colonies can contain several million pairs of
breeding birds and can cover tens of hectares.[2]

The entire breeding cycle, from nesting colony to independent
fledged young, requires six weeks—an exceptionally short interval for
birds. The biological synchrony found among the birds in a breeding
colony can be remarkable. Millions of eggs in millions of nests hatch on
the same day. The fall of eggshells from subsequent dropping of the
shells from the nests after these synchronous hatchings has been
likened to a snowstorm.[3] Sometimes the dry season comes early and
the breeding areas dry out prematurely. In such times the breeding
colonies are abandoned. In other instances, the rains may sustain a pro-
longed green period and several episodes of breeding can occur in the
same general area.

Since red-billed queleas feed on the seeds of annual grasses, they are
preadapted to be effective feeders on the seeds of domesticated annual
grasses as well. As human agriculture has planted more and more cereal
grains across Africa, the numbers of quelea have exploded in response to
the abundance of suitable food. The quelea's magnitude as a pest animal
has increased correspondingly.[4] Although the numbers may be overesti-
mated, the species is certainly capable of destroying 10 percent to 20 percent of the production of large farms and the entire crop of subsistence
farmers.[5] The red-billed quelea is not only the most abundant bird, it can
also be described as the most destructive pest bird on Earth.[6]

Based on the observations at a given location, the arrival of huge
flocks of queleas would seem random. To a subsistence farmer in Ethi-
opia, Somalia, or Kenya, the appearance of red-billed queleas from

someplace beyond the horizon, and the subsequent rapid destruction of his crops by the birds, would inspire a categorization of the quelea as not just a pest but a pestilence.

Red-billed queleas do provide a valuable source of protein in Africa, however. Using a variety of nets, different ethnic groups in Chad and Cameroon trap quelea for food. Hadjerai people working in six-man crews trap and process about 20,000 birds a day. Quelea are plucked, fried, dried in the sun, and sold in markets. The revenues from selling them are something like 40 percent of the value of the crops they destroy.[7]

Like the red-billed quelea, some animals and plants have a remarkable capacity to expand their populations under the appropriate environmental conditions. When landscapes are altered by extreme change, these species can demonstrate remarkably abrupt changes: rare species can become common; superabundant species can disappear. In geologic history, and in the shorter historical record, are numerous examples of such superabundant species.

The potential to respond strongly to fleeting opportunities occurs in plant species as well. Philip Grime, professor of plant ecology at the University of Sheffield, has spent much of his professional life categorizing the manner in which plants respond to their environment (and to changes in that environment).[8] Grime recognized two types of external environmental factors that limit the biomass of plants at a given location. The first is *stress,* which involves conditions that restrict plant productivity. Examples would be a shortage of light, water, or mineral nutrients. The second external factor is *disturbance*, which involves partial or total destruction of plant biomass (through the activity of herbivores, diseases, fire, frosts, and the like).[9]

The previous chapter presented disturbance based on the examples of the Bachman's warbler and the Leadbeater's possum. In Grime's categorization, disturbance and stress are factors that can operate independently of each other. Thus, there are four possible combinations of high or low stress, and high or low disturbance: high stress with high disturbance, high stress with low disturbance, low stress with high disturbance, and low stress with high disturbance. The combined action of high stress and high disturbance creates a condition from which the vegetation cannot regenerate itself and it is not considered in Grime's theory.

According to Grime, low-stress and low-disturbance environments would ultimately favor species that were able to compete effectively against other species (he calls this the competitor strategy). High-stress environments with low disturbance should be dominated by plants able to tolerate the particular stress (stress-tolerator strategy). Low-stress and high-disturbance environments should favor short-lived and fast-growing species (ruderal strategy). These three different strategies (competitor, stress tolerator, ruderal) would have different successes in different disturbance regimes and at different locations. In reality, most plants are seen as being mixtures of the three stereotypical plant strategies.

The red-billed quelea feeds on the seeds of annual grasses, a classic ruderal species in Grime's scheme of plant adaptations. Ruderal plant populations explode in numbers when conditions are favorable. They then disperse to new favorable places or wait for the next opportunity. Ruderal plants grow fast, reproduce early and plentifully, and die quickly. They are the gamblers and nomads of the plant world.[10] These vegetable pirates and botanic mountebanks are not unusual and exotic species from distant locations—we have their seeds for breakfast every time we pour a bowl of cereal. The attributes of ruderal plants are the attributes of the cereal grains that initially were domesticated by humans to develop the agricultural revolution that replaced the hunting-and-gathering style of living.

In natural systems, the adaptations of animals that are specialized in utilizing ruderal plant species tend to parallel the plants' life-history strategies. Ruderal-plant-exploiting animals have high reproductive rates, and production of independent offspring is rapid. Just as ruderal plants feature efficient dispersal to move their seeds to locations with favorable conditions, their seed predators are mobile and move to exploit the plants in new locations. Animal species such as the red-billed quelea, adapted to feed on ruderal plants, are also preadapted to feed on agricultural cereals, the ruderal plants that humans have domesticated. In a sense, queleas were predestined to be agricultural pests.

Farm fields prepared to promote humankind's ruderal crop plants promote other ruderal species of plants as well, which are similarly predestined to become weeds. By human perception, weeds and pests are enemies to be eliminated or at the least controlled. Reciprocally, if weeds and pests could generate comparable evaluations, they would see humans and their agriculture-based promotion of habitat and resources as a boon to their existence.

We have seen that two significant features of the red-billed quelea are its very large numbers and its ability to move to find suitable conditions to survive. To start with the phenomenon of migration, in species such as the red-billed quelea nomadic movement allows the animals to utilize areas that might be unusable or even hostile to survival. In other species, these movements can be much more directed and can be thought of as migrations. The biological adaptations that allow migratory species to navigate vast extents of open ocean to find breeding grounds or feeding areas are truly striking.

In the simplest sense, movement and migration allow species temporarily to avoid harsh conditions. Among the most remarkable feats must be the prodigious flights undertaken by migratory birds. As an example, from their breeding territories across the boreal forest of North America as far west as Alaska (over 1,500 miles or 2,400 km away), blackpoll warblers (*Dendroica striata*) gather in staging areas from New England to as far south as Virginia. Fattening on insects there, they form small flocks that eventually make an 80- to 90-hour, open-ocean flight of 2,200 miles (3,500 km) across the Bermuda Triangle to their wintering grounds in South America.[11] It is an amazing feat for a tiny warbler.[12] When plotted on a conventional map (which flattens and contorts the curved surface of the Earth), a blackpoll warbler's flight from Alaska to New England to Venezuela seems longer than flying to South America by way of the Florida peninsula. Plotted on a globe, the route of the blackpoll warbler is close to a "great circle route," the shortest distance between two points on a sphere. The via-Florida route is actually considerably longer.

The classification of birds as migrants or nonmigrants is not a simple yes-or-no proposition. Rather, it is a continuum of behaviors and adaptations. Some species, such as the red-billed quelea, appear to be quite opportunistic and erratic in their migrations. Other species, such as the blackpoll warbler, are highly programmed and perform improbable feats like flying vast expanses of open-ocean waters or returning from wintering grounds thousands of miles away to the precise location where they were born. Many species of birds do not migrate at all, but are resident in a given region at all times. Presumably, movement over great distances involves risks. Large-scale migration costs energy and produces physiological stress. Thus, one would expect migratory behavior to evolve in cases where the benefits minus the costs of being a migrant is greater than the benefits minus the costs of remaining in place.[13]

Probably the most straightforward form of migration is the irruptive movement seen in species that are periodically stressed by shortage of food or environmental disturbance. Such species can move great distances under these circumstances and can at least temporarily extend their range into new areas. Bird species such as the goshawk (*Accipiter gentilis*) and the snowy owl (*Nyctea scandiaca*) feed on small mammals in the far northern boreal forest and tundra. In years during which there are no prey species (owing to population fluctuations in boreal small mammals and birds), these striking birds of prey invade regions in North America and Eurasia that are thousands of miles farther south—to the great excitement of bird-watchers. Similarly, when the cone-bearing trees that make up the dark conifer boreal forest ringing the upper latitudes of the Northern Hemisphere fail to produce enough seeds, flocks of evening grosbeaks (*Coccothraustes vespertinus*) and red crossbills (*Loxia curvirostra*) spill south in search of food and decorate the bird feeders of suburban homes.

The blackpoll warbler and a great array of other insect-eating small songbirds are obligate migrants. They breed in one region during a favorable period, and they migrate to wintering grounds (which may be very far away) when conditions become unfavorable. Obligate migrants are adapted to take advantage of regions with temporary high productivity where there is a pronounced, and predictable, seasonal variation in environmental conditions.

Between the irruptive and obligate migrants are partial migrants—species composed of some individuals that migrate and others that do not. These populations may involve different genetic traits relative to migration. Or partial migrants may have some segments of the population (younger birds or behaviorally subordinate individuals) that are migratory. In this situation, an individual may be migratory at one time in its life and nonmigratory at others.

One of the central needs of irruptive, partial, or obligate migrants is the capability of finding their way back to the breeding grounds once conditions there have become more favorable. This capability involves timing when to return, a sense of location, and ability to navigate to a specific site. Since such accomplishments can be beyond the scope of cognitive humans armed with maps and compasses, it is hard to understand how small animals with tiny brains can accomplish such feats. Their toolkit includes clocks, compasses, maps, and evolution.

Clocks: Imagine a songbird that has migrated to the Amazon basin for the winter. On the equator, the days are the same length (twelve hours) regardless of the time of year. After six months of identically timed days in a tropical environment with no strong seasonal change, the bird flies north to its breeding ground located in central Canada. If it leaves a fortnight too early, the bird arrives at its breeding areas in a time of inclement weather; a fortnight too late, and other birds of its species have acquired the best mates and the best territories. Clearly, evolution would favor the individuals with the correct solution to this timing problem.

Like a wide range of other organisms, birds have two "clocks" built into their physiology. One is more or less an annual calendar (called circannual rhythm) and the other is more of a daily clock (called circadian rhythm). The daily cycle appears to be reset by timing cues from the environment, such as the light-dark cycle of the day or daily temperature fluctuations.[14] The circannual calendar involves the regular progression of internal physiological cycles that are associated with breeding, molting of feathers, seasonal changes in plumage, and gains and losses of weight. This cycle is maintained even when caged birds are confined in constant conditions with twelve-hour days and twelve-hour nights. One striking aspect programmed into the cycles of obligate migrants is a condition called *Zugenruhe*—a German term for restlessness. In Zugenruhe, birds become more active at night than during the day. This is the condition associated with readiness for migration in obligate migrants.

Compasses: If you were placed in a closed box in a windowless van and driven around the countryside until you were discharged in a remote and unfamiliar forest clearing, you would not expect to be able to immediately indicate the correct direction to your home. Pigeons, however, can.[15] Such trips to distant release points and returns are the basis for the sport of homing pigeons. When released from their transport containers, homing pigeons usually fly immediately in the direction of their home roost. Most experiments in bird homing imply that this remarkable feat has two aspects: a map to determine position and a compass to move in the direction indicated by the map.[16] Birds appear to have four different compasses.

The first compass discovered is the one most familiar to humans. If we know the time (which for birds would involve utilizing their circadian clock), then the position of the sun in the sky indicates direction. The sun is in the east in the morning and the west in the afternoon. Homing pigeons appear to use the "sun compass" to find their way back. But the fact that many of the obligate migrants fly by night implies that they must be equipped with other navigational tools as well.

At night, caged birds in a state of migratory unrest or Zugenruhe will hop on the side of their cages in the direction that they would move if they were free and making a migratory flight. When these birds are moved to a planetarium, experiments on bird navigation can be conducted under artificial starry skies, which can be manipulated. Birds change their hopping direction when the planetarium sky is altered in a way consistent with using the position of the stars as directional cues. It appears that birds learn the stars of the night sky in the first year of their lives and are able to use them as a second compass.

The surprising third compass of birds (as well as of several other animals) is apparently the ability to sense magnetic fields. The lines of force in the Earth's magnetic field are vertical over the magnetic poles and become flatter away from the poles. Using its sense of magnetic field, a bird determines the dip angle or verticality of these lines of force and is thus provided a sense of direction.[17] Migratory birds in Zugenruhe, held in the sorts of experimental cages used in the planetarium experiments, can also be placed in altered magnetic fields using Helmholtz coils. When the magnetic field surrounding the birds is altered, so is the direction of intended movement.[18]

Birds have a fourth compass, the ability to sense polarized light. Humans too have this ability, but rarely know that they do. If you look at a lamp through a polarized sheet of plastic (or even polarized sunglasses), a yellowish hourglass figure appears. It moves if you rotate the plastic. This figure, called Haidinger's brush, points the direction of the polarization of the light. Haidinger's brush is relatively easy to see, but our brains usually edit this phenomenon out of our perception. It is not clear what birds see, but they can detect the pattern of the polarized light that is created by the atmosphere acting on sunlight as a weak polarizing filter. Birds know the position of the sun from patches of sky, even when the sun is obscured from view. Similarly, insects can detect polarized light and use this information as a sun compass.[19]

Maps: Many larger migrants, such as geese and cranes, return with great dependability to the same stopover points during migrations in different years. This consistency indicates a rather precise ability to navigate. Armed with multiple compasses, birds have the information to sense direction under a range of conditions: daytime or night, cloudy or clear. But even with a compass (or multiple compasses), one still needs a map to navigate back to home base. We know less about how this aspect of migration works.

Having a clock, reading the stars, knowing a direction, detecting the dip of the Earth's magnetic fields, and knowing latitude certainly give an individual sufficient information to navigate. However, the calculations are complex. The clocks and senses need to be remarkably accurate to allow birds to migrate with the exactitude that some are known to demonstrate.

Much of the work to understand the map used by birds has been done with homing pigeons. That these birds can return to their lofts even when fitted with frosted contact lenses implies that senses in addition to sight are involved.[20] Birds are now known to have small organs in their heads containing magnetite crystals that can provide a magnetic sense.[21] Pigeons appear to use magnetic fields and magnetic anomalies (places where, for example, large iron-ore deposits disturb the pattern of the Earth's magnetic fields) as maps. To find their way, pigeons, as well as other birds, may use a sense of smell for certain odors blowing in winds from different directions.[22] It has also been proposed that birds can use the inaudible, extremely low-frequency sound called infrasound, which is generated by the Earth's surface, as a kind of map based on sound reflections.[23]

It seems plausible that long-distance migrants could use their sense of direction and their clocks to travel in appropriate directions at appropriate times, and then use magnetic, chemical, or infrasound clues to find their homes.

Evolution: Studies in Europe have indicated that the direction taken by birds on their first migration is inherited genetically.[24] One of the more striking aspects of migration is that the phenomenon can develop (or be lost) from a population in a relatively short period, even in obligate migrants.[25] The story of the packrat (Chapter 4) tells us that the breeding areas that blackpoll warblers now target for their spring return were ice covered 10,000 years ago. Therefore, the complicated behaviors of long-distance migration must have developed since then in this bird. Ten thousand years is relatively short in the scope of evolutionary change, but development and loss of migratory behavior has occurred in short enough periods as to be observed by human observers—in time spans measured in decades rather than millennia.

For example, the nonmigratory house finches (*Carpodacus cassinii*) introduced to Long Island, New York, in the 1940s have now spread across the upper half of the eastern United States. Now a consistent visitor to bird feeders in suburbs of the east, the species has also become a partial migrant. Rufous hummingbirds (*Selasphorus rufus*) from the Pacific Northwest of the United States and Canada that normally migrate to Mexico for the winter have learned to migrate by the hundreds to backyard hummingbird feeders in the southern states of the Gulf Coast. Other migratory species when introduced to new situations have become nonmigratory in equally short periods.

Migration is one principal feature of the life and success of the red-billed quelea. A second feature is the astounding densities that the quelea can attain. The *Guinness Book of World Records* entry quoted at the beginning of this chapter puts the number of quelea at 10 billion—which is 10^9 or one thousand million. If this figure and an earlier estimate of the number of birds in the contiguous United States are correct, there are about twice as many red-billed queleas in Africa as there are birds in America.[26] Remarkable numbers are seen in some other highly mobile migratory species as well.

The quelea is a remarkable successful animal at moving to average out the subcontinental variations in its food supply. It uses movement and biological adaptations for explosive population growth to exploit favorable conditions and to flee unfavorable ones. Migratory animals have evolved complex behaviors with remarkable compasses and maps to allow similar movements, and they live their lives at continental and even global scales. Such mobility is not without cost. Great energy expenditure and exposure to risk are the price of migration. In the face of large-scale change, the ability to live regionally and globally is not without other risks as well.

The Serengeti still presents vistas of migratory wildebeests (*Connochaetes taurinus*) following the seasonal rains and moving from horizon to horizon. Quelea feeding on crops in Africa move in fluid waves that surge across the land as birds in the rear fly over their peers to feed in the fresh territory at the front of the flock. The staggering abundance of these creatures seems to connote invulnerability to human attempts to control them.

History tells us that these seeming invulnerabilities are potentially more fragile than they initially appear. In North America, bison (*Bison bison*) once covered the landscapes of Nebraska just as wildebeests cover the Serengeti. Now-extinct passenger pigeons filled the air in huge flocks that roosted to break limbs from trees. These creatures left American landscapes dotted with place names such as Pigeon Hill and Buffalo Valley. If the red-billed quelea of Africa is a species that has increased under the influence of large-scale human land-use change, then the passenger pigeon of North America is its antithesis.

The scientific name of the passenger pigeon, *Ectopistes migratorius,* is a descriptive, if redundant scientific name for this species. The genus name is from the Greek εκτοπιςτες, or wanderer, so the name means "wandering wanderer." The passenger pigeon came into recorded history when Jacques Cartier saw an "infinite" number of the birds off of Prince Edward Island on July 1, 1534.[27] With the death of Martha, a captive female pigeon that died on September 1, 1914, in the Cincinnati Zoo, today the pigeon exists only in history. The modern reader is incredulous at the accounts of the species provided by a wide number of observers of its flocks and migrations.

The red-billed quelea uses flocking, communal nesting, and migration to exploit the seeds of annual grasses and other locally abundant but unpredictable resources. The passenger pigeon showed a similar suite of adaptations to exploit an equivalently unpredictable resource, its preferred foods of beechnuts and oak acorns. The American beech (*Fagus grandifolia*) and oak (*Quercus*) species have the most edible acorns, or mast; they produce seeds in large amounts only over intervals. The beech produces an abundance of seeds about every other year, and species in the white oak group have a slightly longer masting interval of three to five years. The trees in a given region are synchronized in their masting; thus, any animals that survive primarily on their seeds are subjected to times of feast and famine. The mast years produce more seeds than can be consumed by the seed feeders, which are numerically diminished by food shortages from the previous year. The passenger pigeon solved this dilemma by evolving a strategy of moving over large regions to find food and of flocking to take advantage of the bounty of a region with masting trees. The bird was hugely successful in this regard—at one time it is likely to have formed 25 percent to 40 percent of the total bird population of the United States.[28]

During the fall and winter, huge flocks of passenger pigeons gathered in roosts in the forests or bottomlands of the South. Flights of pigeons to these roosts, shown in newspaper illustrations of the 1870s (Figure 26), darkened the skies. The birds packed into the roosts at such remarkable densities that their weight broke limbs from large trees and even the trees themselves. The ground over several thousands of acres was covered with 4 inches (10 cm) or more of bird droppings.[29] The clamor of the animals was so loud that the discharge of pigeon shooters' guns could not be heard over the din.[30] Abandoned winter roosts were strange places containing dead and broken trees over large areas, fertile soils from the bird droppings, and inoculums of

Figure 26. "Shooting wild pigeons in northern Louisiana." From S. Bennett, Winter Sports in northern Louisiana: Shooting wild pigeons, Illustrated Sporting and Dramatic News *(July 3, 1875); A. W. Schorger,* The Passenger Pigeon: Its Natural History and Extinction *(Norman: University of Oklahoma Press, 1973).*

seeds of plant species eaten by wintering pigeons. The old roosts provided rich farmland already partially cleared.

The pigeons migrated north in the spring. Often, very large flocks arrived to occupy areas with sufficient food. They avoided regions having off years in the masting cycles, so the birds might nest in one area one year and in a different area the next year.

The species nested in great colonies that averaged about 40 square miles (100 km²) but in some instances were much larger.[31] The colonies occurred in forests and had densities of eighty to one hundred nests in each tree. The birds appear to have laid one egg per nest, but they may have produced multiple broods in the southern part of their range. In the northern part, they only made a single nest per year. This implies a rather low reproductive rate for a species that was so phenomenally abundant, but the birds were long-lived. Captured animals lived ten to twenty years (or even more). Presumably, the mass flocking behavior throughout the life cycle of the birds provided protection. Predators would have not been numerous enough to put much of a dent in the huge colonies.

The species arrived at its breeding grounds fairly early and initiated nest building later in the spring. Arrival in 1882 in Wisconsin (a last stronghold of the species) was on February 9, with nesting beginning in mid-April.[32] The breeding of the species was highly synchronized in all its stages. Mating appeared to require the presence of multiple individuals (perhaps a few hundred, although smaller breeding groups have been recorded).[33] Members of the nesting colony would forage for food about 50 miles (80 km) in all directions. The species was a strong flyer—which provided one of its nicknames, the Blue Meteor. A flight speed of 60 miles per hour (100 km/hr) is generally taken as a reasonable estimate of its velocity.[34]

One of the by-products of a species such as the passenger pigeon is the disturbance to the forests. In 1728 William Byrd noted, "In their Travels they make vast Havock among the Acorns and Berries of all Sorts, that they waste whole Forests in a short time, and leave a Famine behind them for most other Creatures."[35] The nestings and roosts would destroy and simultaneously fertilize large tracts of forests. The mass actions of such a creature make it difficult to envision the pre-Columbian eastern forests of North America as the vast equilibrium ecosystem, a point discussed in the previous chapter.

The extinction of the passenger pigeon marks one of the sadder episodes of the settlement of the United States. Its tremendous success may have required a continent of vast wildness for the migratory averaging of masting conditions. The pigeon would have undoubtedly been reduced by agricultural land clearing, regardless of other factors.[36] Many naturalists, including Audubon and Thoreau, made this observation in the 1850s and even earlier. When agriculture pushed west in America in the early nineteenth century, the pigeon flocks in the East were diminished.

With the coming of the railroads, which allowed the pigeons to be harvested and shipped back to cities, the pigeon was exploited heavily and wastefully as a commercial meat source. Large numbers of trapped live birds also were released as targets in shooting contests. Trappers entered breeding colonies to shoot, trap, and net birds for market. The subsequent disruption destroyed nests, eggs, and young. Trains and schooners shipped barrels of the birds to eastern markets. The sudden local abundance of a flock would be seen as a heaven-sent source of hard money in frontier and homesteading communities.

As the species became rarer, prices increased and competition for a dwindling resource became keener among the professional pigeon suppliers. Several states passed game laws that regulated the exploitation of the birds. However, these were largely to mediate conflicting hunting practices (bird netting versus bird shooting), or to control access to nesting area at times that would interfere with harvest of the birds.

The species declined to extinction as a wild species from about the mid-1850s until 1900. By 1909 the American Ornithologists' Union was offering rewards for finding the species in the wild to determine if it was indeed extinct. When Martha died in the Cincinnati Zoo in 1914, the passenger pigeon was gone. The stuffed Martha, now a specimen in the Smithsonian's National Museum of Natural History, carries an epitaph as part of its museum label, "Exterminated. Formerly very abundant throughout a large part of North America. This is the last known individual. It died in captivity in September, 1914."

The demise of the passenger pigeon marked a watershed in bird conservation in America and elsewhere. That a species so abundant and so visible could vanish from the countryside was proof positive that exploitation of wild creatures needed to be controlled. Small stuffed wild birds and the plumages of egrets lost their appeal as decorations for women's hats.[37] At the confluence of the Mississippi and Wisconsin rivers, a plaque at the Wyalusing State Park in Wisconsin reads,

> Dedicated To the Last Wisconsin Passenger Pigeon Shot at Babcock, Sept. 1899
>
> This Species Became Extinct Through the Avarice and Thoughtlessness of Man
>
> Erected By The Wisconsin Society For Ornithology

With such recognition of the role of humans in the destruction of a superabundant species, wildlife protection laws were expanded to protect nongame birds—species not normally hunted for food. International agreements were developed to protect birds that migrated across borders.

The red-billed quelea and the passenger pigeon represent radically different but logical outcomes of life-history strategies involving the use of mobility to exploit sometimes abundant but unpredictable resources. These species respond to change in large-scale regions or subcontinents in the same way that other species respond to localized change. Quelea enjoys remarkable success and survives as a pest animal; the pigeon failed and is extinct.

With change, some species populations explode, the red-billed quelea being one example. In the United States, grackles and blackbirds in the South and cowbirds in the wintering feedlots of the Midwest form winter roosting colonies of millions of birds in response to seeds made available from agriculture. Other very large populations suffer catastrophic declines. Giant colonies of seabirds that migrate to distant islands and breed to tremendous numbers on predator-free atolls are vulnerable to the introduction of a few rats, cats, or dogs—one ship's visit can spell their doom. In altered environments and under human pressures, vast numbers of animals do not necessarily assure their invulnerability to change.

7 The Engineering Rodent

Vancouver, B.C., September 8.—All trains held up on the main line of the Canadian Pacific by slides are now on their way east and west. The line was finally cleared at eleven o'clock this morning. Yesterday the population of Field, the town nearest the slide, which occurred between Pallises and Glenogle, 25 miles west of Field, was increased by the addition of 2,000 passengers who were held up by the delay.

The slide, which was 300 feet wide and 30 feet deep, was caused by the bursting of an old beaver dam high up on the mountains. District Supt. MacKay, at Revelstoke, says that the dam burst under the pressure of the heavy rain storms last week. The slide carried the track away completely and it went clear across the Kicking Horse River, damming that stream and endangering the track above the slide. The river was completely blocked up, and it was found necessary to blast a new channel for the stream to release the pent-up waters which threatened to cause a washout further east.

The Canadian Pacific Railway had two steam shovels and a hundred men at work. Huge trees were brought down with the slide and boulders nearly as big as a box car made the job of clearing the track a difficult one. Some of the trees that came down bore the marks of the little animals' teeth, and the supports of the dam erected by the beavers were plainly marked as such by the bleaching of their upper ends and the lower points coated with mud and slime.—News article in the *Gazette* (Montreal), September 9, 1913

Beavers (*Castor canadensis*) are the largest North American rodents. They attain lengths up to 4 feet (1.3 m) and can weigh more than 75 pounds (30 kg). Compact, rather thickset animals, they have small eyes, short legs, and broad, scaly, flattened tails (Figure 27). Their paddle tails and their large webbed feet are adaptations to aquatic life. Other aquatic adaptations include valves in their ears and nostrils that close when they are submerged. A clear membrane, called the nictitating membrane, covers the eyes when beavers are underwater. Beavers have dense fur: a fine gray underfur overlaid with coarse, darker brown guard hairs. The animals feed on a wide range of plant material, mainly the bark of wetland trees and shrubs.[1]

One of the beaver's outstanding attributes is its capacity to construct complex civil engineering projects. These modifications of small streams can involve dams of mud and sticks to heights of 6 feet (2 m) or more and lengths of up to a third of a mile (500 m). Beavers sometimes construct canal systems for transporting small trees and cuttings. In many instances they build "lodges" in the lakes formed by their dams. A typical beaver lodge is made up of sticks in the middle of a pond

Figure 27. The American beaver, Castor canadensis. *From J. J. Audubon and J. Bachman,* The Quadrupeds of North America, *vol. 1. (New York: V. G. Audubon, 1854).*

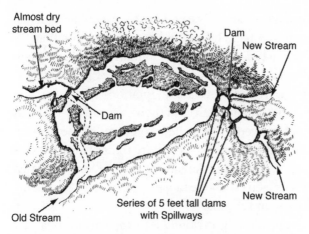

Almost dry
stream bed

Dam
New Stream

Dam

New Stream

Series of 5 feet tall dams
with Spillways

Old Stream

Figure 28. A beaver pond in North Ontario, Canada, with multiple dams and a stream rerouting. From A. R. Dugmore, The Romance of the Beaver *(Philadelphia: J. B. Lippincott, 1914).*

formed by a beaver dam, or on the side of a stream with an associated burrow. Beavers also construct nest burrows in the sides of river or creek banks. The entrance to the lodge or nest is always from below the water's surface.

Like the beaver dam that burst and stopped intercontinental train transport in the quotation from the Montreal *Gazette,* beaver projects can be impressive in their scale and complexity. One example (Figure 28) involved the complete rerouting of a stream, creating a pair of new dammed streams and a cascade of dams with spillways.

The beaver population of North America at the time of European arrival is estimated to have been somewhere between 60 million and 400 million animals.[2] The species was found in virtually all aquatic habitats, particularly streams, from the north Mexican desert to the arctic tundra. The beavers' geographic range encompassed about 6 million square miles (15 million km²).

The species was exploited for its fur from the early seventeenth century until the latter part of the nineteenth century. As beavers fell victim to fashion in the East and were eliminated, explorers—followed by rough-and-tumble frontiersmen, mountain men, and voyageurs— pushed into the western wilderness to find more beaver skins.[3] By 1900 the beaver was almost extinct in North America.[4] Today a managed population continues to recover from historic overexploitation for fur, but the species probably still numbers less than 5 percent of its former density.[5]

Trapping North American beavers for their fur began with the Italian explorer Giovanni Caboto (John Cabot), sailing for the Merchants of Bristol in 1497. Among other findings he discovered the Grand Banks, an extensive set of shallows off the coast of Newfoundland, Canada, where codfish were so numerous that they could be brought up with baskets. Caboto's discovery made relatively little impression on his employers in beef- and mutton-eating Protestant England, but Catholic Europeans found it to be of considerable interest. Six hundred ships were working the Grand Banks by 1603 to produce salted and dried fish for shipment to Europe. Negotiations with several of the tribes of Amerindians on shore involved gifts to ensure that the fish drying racks (an obvious source of easy-to-get firewood) would be left for the next fishing season. These initial commercial contacts eventually propelled an expanding trade in furs, including beaver pelts. Subsequently, Europeans formalized trading relations with a diversity of native people, based on exchange of furs for a wide range of goods. A trade network ultimately stretched across the North American continent.[6]

New England is a microcosm of the expansion of fur trading in North America. Prior to European settlement, beavers were found in almost every body of water.[7] Between 1620 and 1630, an average of 10,000 beaver pelts per year were taken in the fur trade in Connecticut and Massachusetts.[8] Indeed, between 1631 and 1636, returns from the trade in beaver pelts amounting to £10,000 paid the debt of the fledgling Plymouth Colony. In New Netherlands (the New Amsterdam colony and its outreaches), the Dutch West India Company's exports progressed from 400 beaver skins in 1624 to almost 15,000 beaver pelts annually by 1637.[9] A century and a half later, by 1764, very few beavers were being taken in Massachusetts (or elsewhere in New England).[10] As the species was locally wiped out by overtrapping, the westward expansion of trade and trapping commenced.

In its heyday as a hat-making material, beaver fur had been converted to a fine gray felt. The demand was fueled by European male fashion. The exploitation of beaver for hat-felt spread across North America at an ever-escalating pace. Then, in 1840, a hatter's apprentice

in London discovered that chopped silk fibers could be used to produce felt hats with a greater luster than those made of beaver felt.[11]
The silk top hat became the mark of a well-dressed gentleman, and the
price of beaver fur collapsed by 30 percent or 40 percent over the next
fifty years.[12]

Even with the decrease in the demand for beaver pelts caused by the
switch in European hat styles, beaver continued to be exploited as a fur
animal. The Hudson Bay Company sold 3 million beaver skins from
1853 to 1877.[13] By the middle of the nineteenth century, the combination of hunting, trapping, and conversion of wetland habitat had taken
the beaver populations down to a small percentage of their former
abundance. The species seemed destined for extinction.

Today, many of the natural predators of the beaver (notably the gray
wolf, *Canis lupus*) have been reduced or eliminated, and laws regulate
the trapping of the species. Beaver populations are expanding and the
population of North America is thought to be on the order of 6 million
to 12 million individuals, still a small fraction of the former abundance.[14] The loss of such a marvelous creature as a consequence of what
in retrospect seems an almost trivial fashion has a darkly ironic aspect.

In North America the local demise of the beaver was followed by the
clearing of land and agriculture settlement. Exploration and subsequent exploitation of natural resources often tended to move upriver.
Tributaries of a river could feature the elimination of beaver on the upper reaches of a drainage system, while clearing and farming were altering the surrounds of the lower reaches.

In historical accounts, eastern North American waters are quite different from their contemporary condition. Great rivers were clear,
with numerous woody snags, complex backwaters, and forested floodplains.[15] The bottom of coastal bays and lagoons was visible. For example, the Chesapeake Bay was transparent and full of eelgrass beds,
along with extensive shell reefs with oysters of remarkable size. Today
the eelgrass is gone and the muddy bay has a silt bottom that is a difficult substrate upon which to establish oysters.[16] These changes appear
to be strongly related to the increased sediment that was fed into the
bay from poor agricultural practice and subsequent soil erosion.

Sedimentation rates in the Chesapeake before European settlement
were on the order of 1 inch every thirty years (0.09 cm/yr). This rate
doubled when the surrounding land was about 20 percent cleared and

doubled again when the land clearing reached 50 percent.[17] But was the siltation due to other causes, perhaps overexploitation of the Chesapeake Bay's extensive oyster reefs that filtered and refiltered the Chesapeake's water?[18] Could it have been caused by the lack of clean source waters resulting from the loss of beavers and their dams? Perhaps all of these questions can be answered in the affirmative. The crux is really the degree of influence of each of these factors.

The presence of beavers and their projects strongly affects the streams that feed rivers. First-order streams are the smallest ones in a drainage system. A second-order stream is formed when two first-order streams join, a third-order stream is created by the joining of two second-order streams, and so on. With occasional catastrophes such as the one described at the start of this chapter, beavers are able to construct dams on streams up to fourth order. Streams any larger have floodwaters strong enough to regularly wipe out the beavers' construction projects.

The nature of smaller streams is substantially altered by the presence of beavers. The ponds caused by their damming of small streams act as settling basins for silt and sediment. The emergent vegetation in the shallows of the ponds ties up nutrients in the water and serves to reduce what in present times is called stream eutrophication—the overenrichment of the waters. The changes caused by beaver ponds are eventually transmitted downstream to the higher-order reaches of rivers. We have seen that beaver projects have the ability to modify water courses and water quality. The filling of beaver ponds builds rich moist meadows, which go through ecological succession to produce diverse plant and animal communities.[19]

Primarily, the beaver demonstrates the key role of animals in the functioning of landscapes. This short chapter provides a transition between the animal stories that have come before, which are largely involved with the ecological concepts in "natural systems," and the stories that follow, which add the human component of landscape change.

The beaver is an animal agent of landscape change. Its story establishes the need for careful stewardship of the species and of similar creatures that have the capacity to alter the systems in which they occur. Indeed, this point has been elaborated with a rich panoply of studies of the role of individual animals in ecosystems around the world. Discussions of the effects of animals on ecological systems not only fill books, they fill library shelves and sections.

Animals pollinate, bioturbate, predate, and aerate; they ingest, digest, egest, and divest vegetation of seeds, buds, bark, and roots. They are a small part of the living biomass (surely for terrestrial ecosystems), but they can have profound effects on the structure and functioning of the ecosystems in which they occur. Charles Darwin, after traveling the world and developing the theory of evolution, settled into the important problem of understanding how the actions of earthworms modify the soil.[20] Thousands of other biologists and ecologists have dedicated their lifetimes to deriving similar insight into the effects of other creatures on ecological systems.

What emerges from this rich body of work is a "law of unintended consequences." We can find numerous examples of ecological systems that appear susceptible to change. The elimination (or conversely, the introduction) of what might seem a minor species can potentially produce profound and unexpected results in natural landscapes, sometimes over large areas. Some of these unintended consequences have had huge economic impact and have altered human history.

Some ecosystems are changed greatly by the elimination (or addition) of a single species. Other ecosystems are extremely robust with respect to some of their features, even in the face of remarkable levels of species extinction. The deciduous forests of the eastern United States, Europe, and eastern China are very similar in terms of their physical appearance and in the taxonomy of the component species. Further, the interaction between a forest ecosystem and the underlying geology produces soils with similar physical and chemical properties in all three continents. Yet the number of species and genera in Europe is considerably less than in North America, which is still less than in the comparable forests of China.

The European diversity, once much greater, has been regularly eroded by extinctions resulting from the periodic glaciations in the Pleistocene.[21] Over the past 2 million years or so, as glaciers have proceeded and receded across Europe, many species and genera of major tree species have been lost. For the region of Europe that is now The Netherlands, the number of tree genera has dropped from about forty-two to around fourteen today. Some of these "lost" genera are not only gone from The Netherlands, but from all the rest of Europe as well. Species of these same genera are still important trees in North America and Asia. How has the loss of such a substantial section of tree diversity from the forest vegetation of Europe changed the way these ecosystems function? The answer is far from obvious. Yet European data on nutrient cycling, productivity, and water use by forests apply to forests in North America, and vice versa.

The differences in biotic diversity among the three major temperate deciduous forests (China, North America, and Europe) are substantial. Species or genera with seemingly unique and important ecological attributes may be missing in one of these deciduous forests but present in the others. A similar observation can be made for the tropical rain forests of Africa: they are floristically less diverse than those of South America, which in turn have lower diversity than the Southeast Asian rain forests.[22] We are only beginning to understand the consequences of such differences. Perhaps because the period of scientific data collection on forests has been short relative to the slow response of forests, there is a great deal more that we need to know.

Certainly we have seen instances where the elimination of a key species, or the introduction of a new species, has resulted in large and in some cases surprising changes in natural ecosystems. We also know of instances where changes in ecosystems follow the loss of an important species or group of species that carry out some essential function in an ecological system. For purposes of illustration, here are a few examples.

• *Pollinators:* The tremendous explosion in the diversity of flowering plant species some 100 million years ago was probably caused by the interactions between plants and pollinating animals.[23] Some plants are highly dependent on specific pollinators and compete for access to them. The adaptations to secure pollinators with a high fidelity to the flowers of a particular species can be marvelously complex: differences in the shape, size, and position of the flowers; timing of the flowering; and a variety of chemical cues and attractants. One of the most surprising adaptations is the coevolution of biochemical poisons in the nectar of flowers that can only be tolerated by the specific pollinator of that plant.[24] In the most highly developed plant-pollinator coadaptations, if one of the partners is lost, then so is the other. Similar adaptations associated with animals in the dispersal of seeds are also often highly co-evolved and represent an equivalent case of strong interdependency between species.

• *Keystone species:* The sensitivity of highly coevolved pollination or seed-dispersal systems to the extinction of one of the partners represents a particular case of a larger category called keystone species. By analogy to the keystone that holds together the stones in an arch, these are species that hold together the interactions among a group of species.

For example, in the central wheat belt region of Western Australia a collection of nonmigratory birds called honeyeaters (family Meliphagidae) live from flower nectar and pollinate an array of attractively flowered shrubs and small trees in the Proteaceae family. The honeyeaters are attractive, mostly small birds; the plants that they pollinate are to some degree adapted to them.

And many of the region's plants are dependent on the honeyeaters for pollination.[25] The different flowers are of sizes and shapes to accommodate honeyeaters of appropriate sizes and with particular bill lengths. The plants flower at different times, a phenomenon associated with timing reproductions so as not to compete with other flowers for the services of an appropriately-sized honeyeater. However, during one short time of year only one species of *Banksia* (*B. prionotes*) blooms, and it sustains all of the honeyeaters through a critical period. This *Banksia* species is widespread, but it is relatively uncommon and quite locally distributed in the central wheat belt region. If this one plant were somehow eliminated, the honeyeaters and all of the species they pollinate could disappear from the region.[26] Thus, *Banksia* is a keystone species.

Or a keystone species could be a predator that controls a pest which might otherwise destroy the vegetation; it could be a pest or disease that keeps the density of a predator from becoming too great; it could be a necessary food plant. If keystone species are a common part of the fabric of ecological communities, then they are agents for surprise in management of ecological systems.

• *Pests and diseases:* Probably the most easily documented large changes in ecological systems from a single organism result from pests and diseases. Animal pests can be seen as exercising strong control over the success of plants and other animals; they can control the abundance or even the presence of dominant species of plants and animals and thereby change the nature of the ecosystems. A fungus parasite, the chestnut blight (*Endothia parasitica*), was introduced into New York on imported Asiatic chestnut nursery stock.[27] Within thirty years of its discovery (in 1904), this fungus had destroyed virtually all of the American chestnut (*Castanea dentata*) timber in the eastern United States.

The paradox that ecosystems are observably resistant to elimination of species in some cases and highly sensitive to seemingly analogous species deletions in others is a consequence of internal feedbacks. Understanding the basis of animal interactions motivated Charles Elton to introduce an ecological niche concept that emphasizes the way species feed on one another. This concept increasingly stressed the competitive interaction of animals to generate patterns of abundance. Some ecological systems seem to be composed of highly interactive species populations whose interactions can be seen as generators of patterns. This feature seems not to occur in other ecosystems. The interactions among components produce a spectrum of regulatory internal feedbacks, which range from being system altering to being inconsequential. Sadly, we are not exceptionally successful at differentiating major from minor interactions a priori—at least not at present. We do understand ecosystem feedbacks in theory, and we are developing a portfolio of what has happened in the past when environmental change has yielded large ecosystem consequences.

We need to make significantly more discoveries before we can confidently predict the results of eliminating species from ecosystems. Nonetheless, it is clear that plants and animals are not passive players. Large animals can destroy parts of the systems in which they occur, and creatures such as the beaver construct visible changes to the landscape structure. It is certainly no less true of the smaller creatures that make subtle alterations in the fabric of ecosystem interactions. Change in the vectoring of disease, and in the activity of animals as parasites and pests, can profoundly alter the nature of ecosystems. The fungal and bacterial components of ecological systems dominate ecosystems as diseases, processors of material, and transformers of the physical and chemical structure of the soil. In a time of species extinction, the effect on ecosystem functioning of change in the biota must be better understood.

8 The Fall of the Big Bird

Early in 1843 I removed from the Bay of Islands to Wanganui, and my first journey was along the coast of Waimate. As we were resting on the shore near the Waingongoro Stream I noticed the fragment of a bone which reminded me of the one I found at Waiapu. I took it up and asked my natives what it was. They replied "A Moa's bone, what else? Look around and you will see plenty of them." I jumped up, and to my amazement, I found the sandy plain covered with a number of little mounds, entirely composed of Moa bones; it appeared to me to be a regular necropolis of the race.—R. Taylor (1873), reporting on his discoveries in New Zealand

New Zealand broke away from a giant continent of the Southern Hemisphere called Gondwanaland between 80 million and 85 million years ago. Over time, Gondwanaland further fragmented to produce the continents of Antarctica, Africa, Australia (including New Guinea), and South America. Smaller pieces separated to become Madagascar, southern India, and New Caledonia. As the Tasman Sea formed between New Zealand and Australia, New Zealand's fauna and flora were launched on an evolutionary trajectory independent from the rest of the world.

After New Zealand separated from Gondwanaland, any creature unable to cross a considerable distance over ocean water could not spread there. For this reason snakes and most mammals, which evolved after the separation, are not indigenous to New Zealand. Bats, which could fly there, and seals, which could swim there, were the only mammals on New Zealand before the arrival of Polynesian settlers.

When the Polynesians arrived in New Zealand, they encountered birds that had been evolving for 80 million years without the presence of mammalian predators. Among the most striking of these animals must have been the moas (Figure 29). These were gigantic wingless birds standing as much as 10 feet (3 m) tall and weighing as much as 550 pounds (250 kg).[1] They are known from a diverse array of remains including eggshells, eggs, a few mummified carcasses, vast numbers of bones, and some older fossilized bone. The eleven moa species that are currently recognized occupied ecological niches customarily filled elsewhere by large mammalian browsing herbivores. They may have had relatively low reproductive rates; apparently, they usually laid only one egg at a time.[2]

Moas ranked in height from the tallest at about 10 feet to smaller species the size of a large domestic turkey (about 3 feet, or 1 m, tall and weighing 45 pounds, or 20 kg). They were unique in having neither wings nor even residual wing bones. As one expects of large birds that feed on vegetable matter, moas had muscular gizzards. They swallowed small stones up to 2 inches (50 mm) in diameter into their gizzards for grinding food before digestion. These polished stones, called gastroliths, often occur in groups along with moa bones.[3]

Many gastroliths have been found in what are now human-modified grassy habitats, giving the initial impression that moas were grass-eating animals. But the present vegetation at a site may not be its previous vegetation.[4] Based on preserved crop contents from mummified specimens, moas fed on leaves, seeds, and green twigs of trees and parts of shrubs.[5] Thus, it appears that they were creatures of the forest and shrubland—more like browsing deer than grazing cows.

Figure 29. The giant moa, Diornis giganteus. This bird had a height of 10 feet (3 m) and weighed as much as 550 pounds (240 kg). From a drawing by Paul Martinson in B. Gill and P. Martinson, New Zealand's Extinct Birds *(Auckland: Random Century, 1991).*

Moas were not delicate creatures. Sometimes the bones of the larger birds were confused with oxen bones by nonscientist observers, at least on initial inspection. Their size meant that they were more likely than smaller and more delicate birds to leave fossils and semifossil remains. Because they were unique and because they became extinct so quickly, moas and their sad recent history have piqued the interest of ecologists, paleoecologists, and paleontologists.

It seems possible that when Captain James Cook first visited New Zealand in 1769, moas (or at least one of the moa species—the upland moa, *Megalapteryx didinus*) may have still survived in the remote areas in the western part of New Zealand's South Island.[6] If so, these individuals would have been the last of their kind.

The Maori, New Zealand's Polynesians, arrived about a thousand years ago. From about 900 to 600 B.P., they intensively exploited moas for food, feathers, bone, and skin. After this time moa hunting declined, but opportunistic hunting of the animals continued until about 500 B.P. on the east coast of the South Island and in its western interior until as recently as 300 to 200 B.P.[7]

Climatic conditions in New Zealand appear to have been relatively stable over the period during which moas became extinct. Different causes could have worked in concert to account for their abrupt disappearance:[8]

- *Extensive forest clearing:* Vegetation considerably altered in the Polynesian occupation of New Zealand, a change not easily explained by climate variation or other possible factors. Forest and shrubland burning appears to have reduced the prime habitat of many of the moa species. However, the main forest burning started around 700 B.P., after what current archeological evidence indicates was the most intensive stage of moa hunting.[9] While there appears to have been extensive burning on the east side of the South Island of New Zealand, large forest tracts remained in the most southern part of the island. Because major habitat destruction seems to have occurred after the moa populations already were depleted, and because some habitat that could have sheltered populations of moas remained, it would seem that other factors were also at work in the extinction of these birds.
- *Direct human predation:* For the South Island, human predation

appears to have been a significant factor in the depletion of the moa population. From one location alone, six railway carloads of moa bones went to the bone mills at Dunedin.[10] The density of Polynesian settlements and artifacts increased substantially at the time of the most intensive moa hunting (900 to 600 B.P.). This period was followed by a time of decline in the Maori population and a societal transition to smaller, less numerous settlements.[11] The apparent decline fits the pattern expected as a consequence of their over-exploitation of moas.

• *Introduced organisms:* The Maori introduced the Polynesian rat (*Rattus exulans*) and the dog (*Canis familiaris*) to New Zealand. The actions of these potential nest predators could have reduced moa populations without leaving much direct evidence. The Maori inadvertently may have also brought pests and disease organisms in fowls that could have crossed over to eradicate moa populations. The possibility of using ancient DNA to identify past diseases of extinct animals is being explored.[12] However, evidence of such diseases is difficult to determine directly from paleoecological or archeological remains. For these reasons, it is hard to determine the likelihood that introduced disease organisms were a cause of the moa's decline, but they are potentially significant.

While the last of these possible causes remains speculative, definite clues exist for the action of the first two causes.

Calculations of the numbers of moas killed by the Maori versus estimates of the total number of moas for all of New Zealand at any given time indicate heavy exploitation of the populations. Athol Anderson estimates that the number of moas killed by Polynesians, and the numerous bones stored in known archeological sites, are enough to show that hunting alone accounted for the disappearance of the moa.[13] Even relatively modest rates of hunting can significantly reduce the population of an animal with a low rate of reproduction.[14] Other estimates based on computer models of the human and moa populations indicate that extinction of the moas could have occurred in as little as 100 to 160 years.[15]

The story of the moa and its demise raises ecological issues on the vulnerability of species to human-caused changes—including altered vegetative cover of the landscape, change in the physical environment, and modification of the biota by eliminating some species (such as the moa) and introducing others (such as the dog and rat).

The moa story needs to be explored from three aspects: Is the pattern of extinction in New Zealand unusual? Were the Maori unique among humans as producers of widespread extinctions? What are the expected patterns of change in species survival on islands?

The Maori certainly are not unique among Polynesian people in affecting the islands they colonized. The overall pattern of extinction in New Zealand with the arrival of settlers is similar across Polynesia. The eleven species of moas were not the only ones to become extinct in the millennium of Maori inhabitance of New Zealand. Another twenty-one species of birds appear to have become extinct before European arrival.[16] These include the New Zealand pelican (*Pelecanus novaezae-landiae*); the New Zealand swan (*Cygnus sumnerensis*); two flightless geese (*Cnemiornis calcitrans* and *C. gracilis*); a (probably) flightless duck called the Chatham Island duck (*Pachyanas chathamica*)—as well as three other ducks (*Euryanas finschi, Malacorhynchus scarletti,* and *Biziura delatouri*); three large hawks and eagles; three rails, two of them flightless; a flightless coot (*Fulica chathamensis*); the adzebill (*Aptornis atidiformis,* a unique large flightless rail-like bird); and several other species.

These extinct birds had several common characteristics. They were large, often flightless birds that lived in the habitats most changed by the Maori occupancy. In fact, all of the known New Zealand birds weighing over 22 pounds (10 kg) were lost between the Maori arrival and European settlement. About half of the species in New Zealand that weighed between 11 and 22 pounds (5–10 kg) were eliminated, but more than 80 percent of the smaller birds with weights less than 11 pounds (5 kg) are still extant.[17] The habitat pattern of extinction also reflects the parts of the New Zealand landscape that were most significantly modified by human occupation (Figure 30). Birds found in such habitats tended to disappear more readily than those in other habitats. There are distinct patterns in the species biology of the extinctions that occurred in New Zealand: in addition to gigantism, odd attributes for birds—notably flightlessness—are a common theme in the attribute lists of the extinct species.

The patterns of species extinction and survival found in New Zealand are repeated in other histories of Polynesian colonization. These

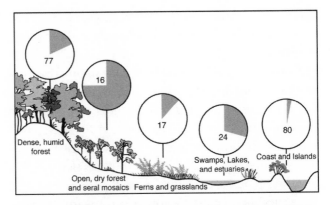

Figure 30. Extinction pattern of birds in New Zealand. The shaded portions of the circles indicate the proportion of the species and subspecies found in each habitat that became extinct in New Zealand during the past millennium. The numbers in the circles indicate the total diversity of species and subspecies associated with each of the habitat types. From A. J. Anderson, Prehistoric Polynesian impact on the New Zealand environment: Te Whenua Hou (pp. 271–283), in P. V. Kirch and T. L. Hunt, eds., Historical Ecology in the Pacific Islands *(New Haven: Yale University Press, 1995).*

patterns have been summarized for several other Pacific islands (among them Chatham Island, Hawaiian Islands, Marquesas, Fiji, New Caledonia, Lord Howe Island, Norfolk Island, and the Kermadec group). As in New Zealand, the extinct species have several frequently shared attributes:[18]

- Extinct species were large birds, often the largest of their kind. This trait was found in both flying and flightless species.
- The flightless species on an island were more likely to become extinct. Those that did were often active by day and were primarily found in forest-edge habitats.
- The extinct species were more likely to have been ground-nesting birds.
- Colonial nesting species were particularly prone to extinction.
- Extinct species were more likely to lay a few large eggs than surviving species, which lay larger clutches of smaller eggs.
- On a given island, the species found only there (endemics) were more likely to become extinct than more widespread species.

The recently extinct birds of the Hawaiian Islands have been well studied compared to those of other islands and provide an excellent comparison to New Zealand.

Storrs Olson and Helen James have summarized twenty years of their work in identifying and classifying bird species from thousands of bones found in sites in the Hawaiian Islands.[19] They described, among other marvels, a collection of highly modified, flightless, goose-like ducks (Figure 31) with turtle-like beaks.[20] These unusual animals appear to have occupied the roles often filled by tortoises (large herbivores with cutting or shearing beaks) on remote islands such as the Galápagos. Olson and James devised a collective common name for these creatures: *moa-nalo.* The neologism comes from *nalo,* which in Hawaiian means lost, vanished, or forgotten, and *moa,* which refers to fowl. In other Polynesian languages, moa carries the sense of a bird and also of edibility. Moa certainly appears to have had the latter connotation in New Zealand. The fruit bat, a favorite food of Polynesians in the Cook Islands, is known as *moa-kirikiri,* meaning leather chicken. The moa-nalo shared the same fate as the moas of New Zealand, vanishing with the colonization of the Hawaiian Islands by voyaging Polynesians in A.D. 300.

The extinct fossil birds of the Hawaiian Islands may number more than sixty species.[21] In the 200 years of naming the birds of Hawaii since the arrival of Europeans, fifty-five living endemic bird species have been described. These described species might number as few as forty if classified by more conservative taxonomists inclined to "lump" species. Olson and James's identification from thousands of bird bones of as many extinct species as the thirty ornithologists who worked with living species over the previous 200 years cannot be overstated; it is simply a remarkable body of work.

Olson and James feel that the length of the roster of extinct birds they have found in Hawaii is clearly a consequence of human action.[22] They state, "Had not *Homo sapiens* arrived in these islands some sixteen centuries ago, these birds would be alive today—skin, feathers, enzymes and all." They implicate direct hunting by people, introduced animals (rats, dogs, and pigs), and possibly introduced diseases. However, they believe that the principal factor in reducing the diversity of birds is the destruction of habitat, particularly in the lowlands where the Polynesians had extensive agriculture when Captain James Cook arrived in 1778.

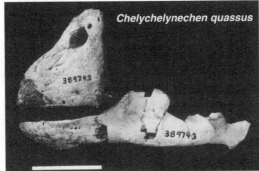

**Skulls of
Hawaiian
Moa–nalo**

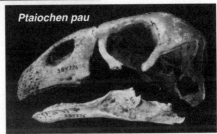

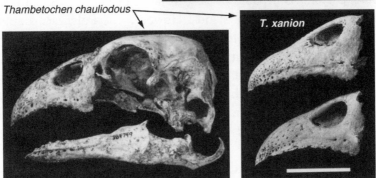

Figure 31. Skulls from the extinct Hawaiian moa-nalo. The white bars in the boxes with the skull fragments of Chelychelynechen quassus and with the beak of Thambetochen xanion are about 1 inch (3 cm) in length. Compiled from S. L. Olson and H. F. James, Descriptions of Thirty-Two New Species of Birds from the Hawaiian Islands: Part I, Non–Passeriformes, Ornithological Monograph no. 45 (Washington, D.C.: American Ornithologists Union, 1991). By permission of the A.U.

Probably in part because Polynesian arrival at an island was an event, it is easier to document archeologically the effect on island faunas and floras. The Polynesians brought with them a domesticated plant and animal package that they used to develop an agricultural support system. This has been called a portmanteau biota, emphasizing the transportability of the plants and animals that sustained the Polynesian agricultural base.[23] Included in this package were several animals (the dog, the pig, the Polynesian rat, and the domestic fowl) and a diverse collection of plants. If we take Hawaii as an example, around thirty-two species of plants were introduced by the Polynesians.[24] Important food plants included coconut (*Cocos nusifera*), breadfruit (*Artpcarpus altilis*), candlenut (*Aleurites molluccana*), taro (*Colocasia esculenta*), sweet potato (*Ipoemoea batatas*), and banana (*Musa acuminata*).

In other islands, these (and several other) food plants were either brought in or were already part of a particular island's flora.[25] For example, coconut is capable of long-distance dispersal and is a native plant on most Polynesian islands. However, the Polynesians apparently brought the coconut to Hawaii. In addition to food crops, plants with a range of other essential uses were either actively transported or utilized from the local flora. Examples are hibiscus (*Hibiscus tiliaceus*), with leaves used for sealing earth ovens; acute (*Broussonetia papyrifera*), with bark that can be pounded into tapa cloth; and ironwood (*Casuarina equisetifolia*), used for war clubs as well as house and boat parts.[26]

Some of the Polynesian islands have been explored by teams of archeologists, paleoecologists, and geologists working together in coordinated studies. These projects reveal the changes in plants and animals on the islands in the context of archeological evidence. They thereby improve our understanding of the effects of human occupation and also of the character of the Pacific islands before the arrival of the Polynesians and, subsequently, after the arrival of the Europeans.

On the island of Mangaia in the Cook Islands, an interdisciplinary investigation determined the changes to the island during its recent history, before and after human habitation.[27] Mangaia is a small, high island (20 square miles, or 52 km^2) with a central volcano. Regional tectonics raised the island, so that what used to be a surrounding reef is now a raised collar around the central volcano (Figure 32). The Mangaians call this raised coral *te maketa*. Makatea (as it is called in English) provides refuge caves and burial places for the Mangaian people. It also serves as a sediment trap for the circle of stream drainages that ring the sides of the central volcano.

At the time of European contact, the Mangaians divided the island into six pie-shaped wedges from the top of the volcano, each including one or more watersheds of the drainages coming down the mountain.

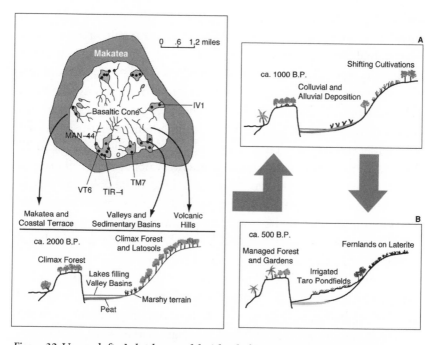

Figure 32. Upper left, *A sketch map of the island of Mangaia in the Cook Islands. Archeological sites are indicated as black dots, and sites used for coordinated studies of vegetation and soils are labeled MAN-44, VT6, TIR-1, and so on. The island has three major vegetation zones with respect to human activities: (1) a center of volcanic hills associated with the cone of the volcano that formed the island, (2) valleys with alluvial soil, and (3) the surrounding makatea (formed from coral platforms raised by geologic processes).* Lower left, *The vegetation pattern as it was thought to have been around 2000 B.P. (before human alteration).* A, *The landscape about 1000 years ago, when forest was still the principal vegetation in the uplands and when the initial erosion of the forest soils had begun.* B, *The landscape pattern about 500 years ago. From P. V. Kirch, Changing landscapes and sociopolitical evolution in Mangaia, Central Polynesia (pp. 147–165), in P. V. Kirch and T. L. Hunt, eds.,* Historical Ecology in the Pacific Islands *(New Haven: Yale University Press, 1995). Copyright Yale University Press.*

A chief (*pava*) governed each of the wedges. A paramount chief (*te mangaia*), whose power had to be legitimized by success in war and human sacrifices, governed the island as a whole.[28] Territories were maintained and defended by the six Mangaian groups in an ongoing internecine war. Each wedge contained all of the principal landscape elements that constitute the island: the basaltic central cone of the island's volcano; the volcanic hills, the valleys, and the sedimentary basins; and the makatea and coastal terraces.

About 1,000 years ago, the initial pattern of settlement and modification of the Mangaian landscape involved modification of the upland forest by clearing and applying a form of shifting cultivation (Figure 32*A*). Small patches of forest were cleared and planted to crops. These small plots were allowed to return to a forest condition as the soil fertility became depleted locally and the crop productivity lessened. When the human population densities became higher, the time for recovery of the cropped patches was shortened and eventually caused substantial soil erosion. The eroded soil from the hillside began to fill the area behind the makatea with sedimentary deposits. A soil process known as laterization converted what remained of the upland soil to laterite, soil in which high rainfall has washed out most of the more soluble compounds, leaving iron and aluminum oxides. When dried, laterite more or less resembles brick. With soil laterization, agricultural production largely ceased in the upland zone of the island.

Five hundred years later (Figure 32*B*), in 500 B.P., human habitation had produced an island whose character was considerably altered from its original state. The lateritic soils of the uplands were covered with ferns rather than forests. The Mangaians had expended considerable effort on working the deposits of soil eroded from the higher elevations into sedimentary basins behind the makatea. As in other parts of Polynesia, complicated hydraulic systems flooded and drained highly productive fields of taro. On Mangaia, the island's pattern of change as a whole is a logical consequence of the interactions of humans, crops, the island's vegetation, and soil. The evidence indicates overwhelmingly that the actions of the Polynesian colonizers altered the landscape of this island substantially.

The changes reconstructed for Mangaia, and the kinds of landscapes produced there, are found throughout the South Pacific. Charles Darwin on the HMS *Beagle* noted the fern-covered uplands of

Polynesia and likened them to the heathlands of Britain. The highly engineered and remarkably productive taro systems of the Polynesian islands impressed many historical observers.

One of the legacies of the Age of Enlightenment is surprise at accounts of the Polynesians eradicating moas and other creatures, seemingly on every island upon which they landed. An enduring myth of the Europeans holds that the Polynesian people, on first contact with Europeans, were the personification of natural nobility. Jean-Jacques Rousseau's (1712–1778) concepts of "Liberty" and the "Social Contract" may have met their ultimate realization in the philosophy of the American Declaration of Independence and the French Revolution. The products of his intellect have profoundly shaped modern history. Certainly one of his enduring ideas was that of the "Noble Savage"— that people in their natural state are intrinsically good.[29] Buoyed by this philosophy, Louis de Bougainville, after his arrival in Tahiti in April 1768, reported on the Tahitians as follows: "I never saw men better made . . . I thought I was transported to the Garden of Eden."[30] Perhaps because of their historical role as icons of intrinsic nobility for idealistic Europeans, we expect the Polynesians to be wise in their stewardship of the land. Indeed, in New Zealand just a few years ago (and even today), the Maori were discussed as "natural conservationists" and "are believed to have lived in harmony with the region and to have altered its character little if at all."[31]

Certainly, the Polynesians are an exceptional people. Their accomplishments as explorers and colonists are astounding—all the more so considering that they are a Stone Age people. Nevertheless, a strong body of evidence indicates that the Polynesians had a considerable and deleterious effect on the environment and biotic diversity of the colonized islands. Were they unique in this regard? Clearly, they were not.

If the Maori colonization of New Zealand caused the extinction of thirty-two species of birds in about 800 years, the Europeans have seen nine additional species lost since they colonized New Zealand about 200 years ago.[32] On a per-century basis, these numbers are roughly comparable with extinctions in European times, among the saddest of which was the Stephens Island wren. *Traversia lyalli* was a nearly or completely flightless little brown bird that ran about more or less like a small mouse. If it was flightless, it would have been the only perching (passerine) bird ever known to be so. The Stephens Island lighthouse keeper was the only European known to have seen the species alive.[33] His house cat single-handedly eradicated the wren.

As noted in Chapter 4, increased hunting pressure from prehistoric peoples may have extinguished the Pleistocene large animals.[34] Similarly, the increase in frequency of fire, by either accidental or intentional means, has been implicated in altering the character of relatively large areas of land.[35] In Europe, considerable prehistoric alteration of vegetation occurred during Neolithic times between 6000 and 4500 B.P., presumably associated with increased human densities and developing agricultural technologies.[36] The conversion of deciduous forests to pine forests in North America at about 5000 B.P. has been tied to archaic Indians setting fires to drive game and to encourage the quality of herbivore browse. Similar practices are speculated to have occurred in Australia as far back as 40,000 B.P. or even earlier.[37]

Certainly, early agricultural societies as well as modern industrial societies are not less significant forces of extinction than the actions of Neolithic peoples. And recent human actions are every bit as detrimental as those of earlier human societies. In modern Polynesia, today's rate of extinction of the surviving species is not fundamentally different from the extinction rates of the past. The technologies are different (guns instead of snares; chain saws instead of stone axes), but fire is used as a tool for clearing land nowadays just as in the past.[38] This appears to be the case not just in Polynesia, but worldwide.

The high rates of species extinction associated with Polynesian colonization and the subsequent introduction of Western technologies reflect a heightened vulnerability to change of island floras and faunas. What makes island species more vulnerable to extinction? To consider this question, we must differentiate remote islands from islands that are relatively close to continents (Figure 33).

The definitive boundaries and isolation of islands have inspired ecologists to look to them as natural experiments in the consequences of landscape size and species diversity. As illustrated by the ivory-billed woodpecker in Chapter 2, the shrinking of potential habitat can be expected to have dire consequences for a species. Can lessons from the species patterns on islands serve to elucidate the consequences of habitat fragmentation on continents? Is the reduction of habitat to patches in a matrix of agricultural land bisected by highways at all analogous to what happens on islands? With a given amount of money to acquire land, is it better to have one large park to protect species diversity or should one develop an archipelago of small parks? Some think that the biogeography of islands holds a key to these questions, but the topic is hotly debated.

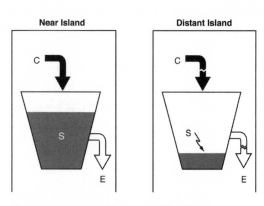

Figure 33. Species richness on an island as a consequence of its remoteness. In classic island biogeography theory, there is a flow of colonizing species onto the island (C). The number of species on the island (S) is a consequence of the balance between the species gained from colonization and the species lost due to extinction (E). On distant islands, the rates of colonization and extinction are extremely low.

In the penguin story of Chapter 3, ecological scale separated and identified environmental factors controlling the distribution of plants and animals. We saw that factors that determine why a species is likely to be found in your backyard and not your neighbor's could be quite different from factors that determine why a species is found in the state of Texas but not in adjacent Louisiana.

For biodiversity of an island, a similar scale concept exists. The factors that explain the diversity of species on an island near the mainland, created by sea-level rise in the past 12,000 years, are not the explanatory factors for island diversity of a distant island removed from the mainland for tens of millions of years—the more so for a remote oceanic island that has never been near a continental source of species. Some islands are so close to the mainland that colonization by new species is a fairly common event; other islands are so remote that colonization by a new species is extremely rare.

"Near islands" are relatively close to continental landmasses and typically have plant and animal species that are similar to those on the adjacent mainland. The islands of the Indonesian archipelago constitute an example of a set of near islands. Many were part of mainland Asia or Australia in the Pleistocene. They originated from the flooding around these landmasses due to sea-level rise at the end of the last ice age.

In contrast to near islands, "distant islands," or oceanic islands, contain mainly endemic species that are unique to either the particular island or the island archipelago. Successful colonization from the mainland is rare. In an archipelago, new mainland species may evolve on one island and then colonize nearby islands. Thus, single-species colonization to a distant archipelago can multiply to several related species over evolutionary time. The islands of Hawaii, the most remote island group in the world, are a classic example of distant islands.

Near Islands

The dynamics of the plants and animals of near islands inspired a concept of diversity called the Theory of Island Biogeography.[39] The essential idea is that the number of species on an island is a consequence of the input of new colonizing species minus the number of established species on the island that become extinct. As it was initially presented, the theory used a relatively simple graphic model of the balance between colonization and extinction of species. The rate of extinction of the species on an island appeared to be related to the size of the island. As was discussed in the Bachman's warbler story (Chapter 5), such an extinction pattern could emerge from dynamic change in the habitat mosaics of islands of different sizes. Since the number of species on an island is the balance between the rate of colonizing species and the rate of extinction, larger islands with lower rates of extinction should support greater diversities of plants and animals than smaller islands.

The classic island biogeography model for near islands is what is called an equilibrium model. Equilibrium, with respect to the number of species on an island, occurs when the input of colonizing species equals the loss of species due to extinction. The general model predicts the number of species on an island based on the distance of the island from the mainland and the island's size. One rule of thumb in biogeography, called Darlington's Law, is that the number of species on islands doubles for every ten-fold increase in island size.[40] This doubling is a direct result of the influence of island size on extinction of species in the island biogeography model.

In the model, the number of species on an island is somewhat analogous to the amount of water in a bucket with a hole in it; the content of the bucket is a consequence of how fast one pours water into the bucket and how fast the water runs out. By this analogy, the more rapidly species pour onto an island (colonization rate) or the more slowly they drain off (extinction rate), the greater the island's biodiversity. Over time, species on the island will be lost to extinction and replaced by colonization. Thus, the model implies a standard length of time that a species, on average, should be resident on a particular island. Over some time interval, the species composition on the island will turn over, or completely change.[41] The rate at which this process occurs is called the species turnover rate and depends on the conditions of the particular island.

The rate that species colonize a near island is directly related to the proximity to a mainland source of species. The closer an island is to the mainland, the shorter the distance that a potential colonizer must travel. Clearly, the frequency of successful colonization depends on what sorts of organisms are involved. Some types of animals and plants are better adapted than others to disperse to distant places.

In the classic island biogeography model, as the number of species on an island increases, the colonization rate is thought to slow. This is a logical consequence of competition with one or more established species acting to limit the growth of the colonizing population. The actual occurrence of a species-related reduction in colonization rate has been the topic of hot debate among ecologists.[42] The scientific literature on island biogeography teems with argument—probably because many of the interpretations are based on scientific inference from a relatively sparse set of actual observations. Nonetheless, understanding the biogeography of islands is central to the conservation of biotic diversity.

We could argue that as the number of species on an island increases, so also should the rate of colonization. Each additional species might help a potential colonist by making it more likely that a colonizing plant will find a pollinator or that a predator will find a prey. The seemingly simple question of whether and why the colonization rate might change with the number of species reflects our central question, "What interactions structure ecological communities?"

We have relatively little direct data on the colonization rates of islands. We have even less information on how frequently potential colonists of islands are unsuccessful. Studies on newly created islands, or on volcanic islands that have exploded, characterize island colonization as a complex process involving soil development from bare surface and the establishment of plants of different life forms (grasses, shrubs, trees, and the like) to generate vegetation and habitat structure.[43] One example is the colonization of the island of Rakata in the Indonesian archipelago.

In the Sunda Strait of Indonesia, a drowned caldera left by an ancient volcanic explosion resulted in the emergence of three islands.[44] The volcano that formed the largest of the three islands, Krakatau, exploded violently on August 27, 1883, after about three months of activity. Huge waves (tsunamis) generated by the blast drowned 36,000 people along the coasts on the nearby islands of Java and Sumatra. Krakatau lost about two thirds of its volume in the explosion. Vast amounts of dust and ash were thrown into the atmosphere, producing meteorological and climatological effects on a global scale. All three islands in the group were stripped of their vegetation by the explosion. Subsequently, volcanic ash between 200 and 260 feet (60–80 m) in depth covered the islands. With the destruction of Krakatau, Rakata became the largest island in the group. Since 1883 Rakata has been the topic of several studies. Until recently, these focused on the plants rather than the animals on the island.[45]

A series of botanists surveyed the island when the pyroclastics subsided. In 1883, they reported the island as devoid of life; by May 1884, the only sign of life was a spider. "Blades of grass," the first evidence of plant life, were recorded in September 1884.[46] The pattern of subsequent colonization was not random; Rakata recovered successionally (Figure 34). Sea-dispersed species initially colonized the strand line of the beach at the rate of over 1.5 species per year.[47] This rate of colonization dropped during the subsequent development of the island. Initially, wind-dispersed species, notably ferns, became established in the interior of the island.[48] The colonization of fern species dropped and then increased greatly when a forest took hold and provided sites for epiphytic ferns. The presence of a forest also allowed epiphytic orchids to colonize. Animal-dispersed plants were not very successful in the initial stage of island revegetation.[49] They became more abundant as more complex vegetation developed in the interior of the island. Each plant type had a different pattern of change in colonization rate as the island vegetation developed.[50]

Although the classic island biogeography model predicts a slowing of the colonization rate as the number of species increases on the island

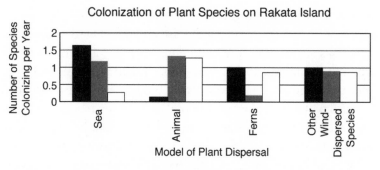

Figure 34. Patterns in the colonization of the Indonesian island Rakata, following the volcanic explosion of the nearby island Krakatau in 1883. Phase 1 is the initial colonization; phase 2 is the establishment of grasslands; phase 3 is the development of forests. Modified from R. J. Whitaker, Island Biogeography *(Oxford University Press, 1998).*

(Figure 33), only the sea-dispersed species actually show this pattern on Rakata. Animal-dispersed species increase in their rate of colonization as the island becomes more diverse. Wind-dispersed ferns initially have a high rate of colonization. This rate decreases and then increases again (Figure 34) as the forest canopy develops and the structure to support wind-dispersed ferns and orchids becomes available. For different species with three modes of dispersal, the colonization rate over time goes up, or it goes down, or it goes up and down. A mixed pattern in colonization rates is found for other taxa (birds and butterflies, for instance). The overall colonization pattern is more complex than the pattern suggested by the classic island biogeography theory.

Since some of the near islands considered in developing island biogeography theory were formerly parts of mainlands flooded by the rise in sea level at the end of the Pleistocene, the idea that the diversity of islands is supplied by colonizing species tends to overlook the diversity of species at the island's starting point. The Sumatran rhinoceros and the Sumatran tiger did not swim the Strait of Malacca to the Indonesian island of Sumatra. The tiger and the rhinoceros were already there when the seas rose to make Sumatra an island. Modern ecologists are studying islands at a time in which the seas are relatively full. In the 10,000 years following the end of the Pleistocene, the seas stood as much as 420 feet (130 m) lower than they do today. Since then, the melting of the continental glaciers has filled the oceans.[51] Along with other geologic changes involving the rise and fall of land surfaces, sea-level rise has surrounded the continents with islands that were connected to the mainland 10,000 years ago. Climatic changes and associated oceanic fluctuations in the recent geologic past make it difficult to understand patterns of plant and animal diversity on near islands. Having only recently become islands, near islands might be expected still to be in a state of considerable biotic change. Unfortunately, we may be applying equilibrium-based models to near-island systems that are nowhere near their eventual equilibria.

Some ecologists have made the argument that even if islands were at equilibrium in the past, the effects of humans on extinctions and colonization have disequilibrated them, so the classic island biogeography model does not necessarily apply.[52] Indeed, the island biogeography model has been criticized with a litany of limitations:

- It considers individual species to be more or less interchangeable and thus ignores the sometimes unique aspects of a particular species.[53]
- Species turnover on islands is difficult to determine. Long-term lists of the changes in the island biota often do not exist and are problematic to create.[54]
- The model lacks precision of definition, which makes it hard to design adequate tests of the theory.[55]
- It is not clear that islands continue to gain and lose species, once a certain diversity of species occurs.[56]

One role of a model is as a simple explanation of a complex phenomenon, offering a better understanding of how the system functions. The island biogeography model has been highly successful at this level, but it has garnered several contraindications to its use as a single basis for conservation ecology.[57]

Distant Islands

Oceanic islands, such as the Hawaiian Islands or the Galápagos Islands, have never been connected to the mainland continents. Other islands—New Zealand, New Caledonia, Madagascar—are geologically derived from fragments of continents, but have been long separated by processes related to plate tectonics and continental drift. These old and remote islands often take on attributes that are quite different from the patterns seen in more recently derived or less remote islands. While the number of species found on near islands may arise from a balance of colonizing species flowing onto an island and extinct species flowing off the island (Figure 33), the model for distant islands is considerably different.

On distant islands, the number of native species that have evolved often is much greater than the small number of successful colonists. Relatively few colonists give rise to large numbers of native species. In Hawaii, for example, extreme cases are seen in the tree crickets (*Oecanthinae*), with as few as four colonizing species radiating into fifty-four species (about 43 percent of the total number of species on Earth), and fruit flies (*Drosophila* and the related Hawaiian genus, *Scaptomyza*), with one or two colonizing species evolving into some 600 or more species—possibly as many as 1,000 if all the species are described. The number of species on the islands is not so much a product of the flows of colonists onto the islands as it is a product of local evolutionary processes and of exchanges of species within the island archipelago (Table 2).

This contrast in near islands and distant islands is related to their ecological and evolutionary scales. In a near island, enough colonization takes place to contribute significant numbers of species to the is-

Table 2. The number of colonizing species and the number of species derived from these initial colonists of the Hawaiian Islands

Animal or plant group	Estimated number of colonizing species	Estimated number of native Hawaiian species
Marine algae	?	420
Ferns	114	145
Mosses	25	33
Flowering plants	272	ca. 1,000
Terrestrial mollusks	24 to 34	ca. 1,000
Marine mollusks	?	ca. 1,000
Insects	230 to 255	ca. 5,000
Mammals	2	2
Birds	ca. 25	ca. 135

Source: S. H. Sohmer and R. Gustafson, *Plants and Flowers of Hawaii*, ed. 3 (Honolulu: University of Hawaii Press, 1993).

land diversity. On a distant island, the evolutionary processes occurring on the island or from species exchanges with nearby islands strongly influence the biotic diversity.

In the colonization of Rakata, different kinds of plants arrived on the island at rates of one or more species per year, and the diversity of the island was strongly related to the colonization and establishment of species. Their evolution was insignificant for species diversification, because of the short time since the Krakatau explosion. In the Hawaiian Islands, the natural rate of colonization is on the order of one species (or even much less) per millennium. Evolution, therefore, is a strong consideration in the nature and diversity of these species.

Distant islands have unique plants and animals that are marvelously adapted to conditions on the islands. These species can be extremely different from the related mainland species from which they were derived. Distant island species tend to share several common patterns of differences from their mainland relatives:

- Loss of dispersal ability: Plants and animals that get to distant islands are likely to have specializations involving dispersal. In plants, the seeds of island species tend to lose barbs, hairs, and other features that aid dispersal. The seeds also tend to be larger. For birds on islands without predators, there is a tendency to develop reduced flight or to become flightless.
- Gigantism and nanism: Island animals tend to evolve in different directions from their mainland ancestors. Individuals of small species become larger (gigantism), and those of large species become smaller (nanism).[58] In the now-extinct fauna of the islands of the Mediterranean are striking examples of both, with giant rodents and dwarf elephants being found. Perhaps an extreme example is the extinct elephant (*Elephas falconeri*) from the island of Malta, which was slightly more than 3 feet (1 m) in height.[59]
- Ecological release: Island species often diversify into a wider range of roles than those of their mainland counterparts. Finches on the Galápagos Islands have diversified into species with many bill sizes (implying different feeding habits) and feeding behaviors (including one species that uses long thorns to probe surfaces in the fashion of woodpeckers). Herbaceous plants on islands often evolve toward tree-like forms.

These and a range of other, more case-specific adaptive differences in the species found on islands make the flora and fauna of distant islands strikingly different from those of the mainlands. For this reason, the biota of islands has been a central topic in understanding evolutionary processes since Darwin and Wallace first developed evolution as a biological theory for the origination of species.

One feature of the biota of distant islands is their apparent vulnerability to environmental change, particularly habitat alternation and the introduction of novel plants and animals.[60] Since 1600, about 80 percent of the animal species known to have become extinct have been island animals.[61] The multiple changes that come with the human colonization of islands are sufficient to produce significant numbers of extinctions on islands—especially distant islands.

Polynesians developed a technology and knowledge base that allowed them to spread to distant ocean islands with ecologically vulnerable biotas. The Maori, in their interaction with moas and other species in the New Zealand fauna, demonstrate the vulnerability of the biota of distant islands. The rate of extinction on islands colonized by other human societies has been high—certainly comparable to the effects of the voyaging Polynesians on the islands of the Pacific. It is not at all difficult to find examples of the scenario, "Humans arrive on an island; whole-scale extinction occurs." The difficulty is in finding examples where this is *not* the case. Readers with the least inclination to see the Polynesians as unique in their extirpation of fauna need only reread accounts of the passenger pigeon and its demise (Chapter 6) to be dissuaded of this notion.

The factors that have produced extinction and environmental change on islands occur on continents as well. Habitat changes associated with the domestication of plants and animals, the encouragement of some endemic species, and the introduction of exotic species are all agents of change that can alter the structure of the landscape systems of both continents and islands.

9 The Wolf That Was Woman's Best Friend

I noticed a young woman walking along the street, and at the same time suckling several puppies that were wrapped up in a piece of tapa cloth hanging round her shoulder and breasts. This custom of suckling dogs is common to the natives of the Sandwich Islands. These animals are held by them in great estimation, little inferior to their own offspring, and my journeys to the woods often afforded me the opportunity of being an eye witness to this habit. I often saw them feeding the young pigs and dogs with the poi made from the taro root, in the same way as a mother would her child.—J. Macrae, reporting his observations from Hawaii in 1825

The startling observation (startling at least to male European explorers) of women nursing wild animals was not just a curiosity seen in Hawaiian Polynesia. Early chroniclers of such geographically separated people as North and South American Amerindians, Australian Aborigines, and New Guinean Melanesians, to name but a few examples, noted the same practice regularly.[1] One can easily appreciate that such close nurturing of baby animals would reduce their fear of humans, behaviorally imprint them on humans, and ultimately serve to tame them.

Hunting people often encounter baby animals in their pursuit of food and bring them back to a base camp. The practice of nursing and feeding these babies, and the inevitable play with young children afterward, seems to be widespread as a method of taming wild animals for companionship and food. One of the creatures most regularly mentioned as a recipient of succor and affection from women is the gray wolf (*Canis lupus*) and its domesticated descendant, the dog (*C. familiaris*).

Dogs are familiar—and remarkable—animals. Most breeds appear to have a relatively recent origin.[2] They are the most variable mammal species. Dogs can differ in size from diminutive "toy" breeds to Great Danes and mastiffs. They also vary in their body shapes, length of limbs, and faces. The enormous variability of dogs has inspired some scientists to propose that they are a complex hybridization of gray wolf, coyote (*Canis latrans*), Ethiopian wolf (*C. simensis*), and/or golden jackal (*C. aureus*). All of these species will interbreed with the present-day dog.[3] However, most scholars of dog domestication see the dog as deriving from the gray wolf, in a process that may have occurred several times.[4] The possibility that domestication occurred independently among the different subspecies of wolf may be a source of their variability.[5] Many of these subspecies, particularly those that are Asian, are quite a bit smaller than the more robust wolves of the far north that invariably come to mind as the standard "wolf."

Dogs appear to have been domesticated before the development of agriculture, at a time when human survival was based on hunting and gathering. Dogs would have been valuable partners in those enterprises. Relationships between Australian Aborigines and the Australian dingoes illustrate the interactions between humans and dogs in a hunter-gatherer society. Like any specific case, this example has its own idiosyncratic aspects. Dingoes (Figure 35) have a complex and, in some aspects, uncertain history.[6] They did not come to Australia at the same time as the ancestors of today's aboriginal inhabitants. Human habitation in Australia dates from 40,000 to 60,000 years ago, possibly even earlier.

The dingo is a relative newcomer, introduced into Australia about 4,000 years ago.[7] At the time in prehistory from which dingo remains date, a widespread change occurred in aboriginal weapons (the use of small spear points), and the largest marsupial carnivores, *Thylacinus* (called the Tasmanian wolf or Tasmanian tiger) and *Sarcophilus* (the Tasmanian devil) disappeared from mainland Australia. The dingo is a logical competitor to these two species, both of which were found only on dingo-free Tasmania once the dingo became established on main-

Figure 35. The Dingo, Canis lupus dingo. *Photograph from the Australian Commonwealth Scientific and Industrial Research Organization (CSIRO), Division of Wildlife and Ecology. Reproduced by permission of CSIRO Australia © CSIRO.*

land Australia. The dingo never reached Tasmania, which by 11,000 B.P. was disconnected from mainland Australia by sea-level rise. Thus, the dingo introduction occurred *after* the great continental glaciers of the Pleistocene had melted to fill the seas and separate New Guinea (along with several other islands in what is now Indonesia) from Australia.

Almost without question, dingoes came to Australia as passengers on boats from some location in southern Asia. They physically resemble South Indian pariah dogs and Thai dogs.[8] Currently classified taxonomically as *Canis lupus dingo,* indicating that they are subspecies of the wolf, dingoes are different from dogs (even dogs from Australia that look more or less like dingoes) with respect to the shape of their skulls.[9] They overlap in these same skull measurements with wolves—particularly Asian wolf subspecies (the Indian wolf, *C. lupus pallipes,* and the Arab wolf, *C. lupus arabs*).[10] If the voyaging Polynesians, a later group of seafarers who may have been descendants of the suppliers of the dingo to Australia, can be taken as an example in the use of their dogs, the dingo may have been on board as a portable food item for the voyage to Australia.[11]

In its interactions with Aborigines, the dingo reflects many of the expected features of the taming and domestication of wild animals. Aboriginal women nurse and tame dingo pups removed from the wild. Dingoes are notoriously difficult to train, except when captured as very young pups with their eyes still closed.[12]

Observations from the Gibson Desert in Western Australia reported dingoes living on what they could procure on their own and rarely being fed by the Aborigines.[13] At the same time, these Aborigines often petted and groomed their dingoes with great affection. Because hunting in this particular tribe was by concealment, dingoes were driven away from hunting parties and were considered a nuisance in obtaining game. Their principal use was to huddle together with people for mutual warmth during the near-freezing desert nights.[14] A "three-dog night" is a night cold enough to require three dingoes to stay warm. The radiative loss of heat to the cold night sky in the clear air of a desert is not trivial; the dingo's role as a blanket could be very important indeed.

Other aboriginal peoples do employ dingoes for hunting. As two examples, the Garawa tribe (located near the Gulf of Carpentaria in northern Australia) uses dingoes to chase down wounded game, and the Mildjingi tribe (from Arnheim Land) makes similar use of them.[15] Dingoes and domestic dogs are active in the sanitary cleaning of present-day aboriginal camps—a function they may have performed since antiquity.[16]

Dogs were the first animals domesticated by humans, probably more than 14,000 years ago. This time in prehistory is indicated by three rather different lines of analysis. First, several examples of fossil dog bones have been found in conjunction with human settlements. Probably the earliest dates for fossil dogs are from Germany, 14,000 years ago, and from Israel and a cave in Iraq thought to be more than 12,000 years old.[17] Other finds of dogs, from widely dispersed locations including North America, are only slightly later.[18]

A second line of evidence on the time of origin of dogs derives from DNA analysis of the variations in soluble proteins among different species of Canidae, the family of carnivores that includes wolves, coyotes, and foxes. The use of DNA to identify the blood, semen, or saliva of perpetrators in criminal proceedings is commonly reported nowadays. That the DNA of an individual can be used as a sort of biochemical fingerprint and that related people have similar DNA has moved from being a research topic to being a part of our general understanding. Another application of DNA technology determines the relatedness of animal species.

Mutations cause variations in the DNA or the genes of animals and plants. These mutations are thought to occur randomly and to accumulate in subsequent generations; they show up as variations in the chemistry of an animal's proteins and can be determined from a drop of blood or a small tissue sample. Because the genetic mutations are random and accumulate over generations, two related species become progressively more different with respect to their proteins. Thus, the number of differences between similar species in a list of proteins provides an evolutionary "clock," measuring how long the species have ceased to be in common, interbreeding populations. The greater the number of mutations (measured as differences in the proteins), the longer the time since two species derived from a common ancestor.

Knowing how long species have been separated by means of a genetic-mutation clock provides a map of the evolutionary relations among the set of species that are related to wolves and dogs (Figure 36). By such protein analyses, the domestic dog appears to have been derived from the wolf about 15,000 years ago. Allowing for the range of variation in the calculations, this estimate is quite consistent with the date inferred from archeological evidence.[19]

Further insight comes from a third line of evidence involving the analysis of mitochondrial DNA from a diverse range of dogs and wolves. Mitochondria are components inside cells that are involved with cellular energy metabolism. Like the nucleus of the cell, they also contain DNA. When an egg cell is fertilized, the initial cell nucleus of the new organism obtains half its nuclear DNA from its mother and half from its father. However, the mitochondria in the cells of the new organism are *only* from the egg cell, so that mitochondrial DNA is passed from generation to generation only via the mother.

The accumulations of natural mutations that occur in the mitochondrial DNA provide a second genetic clock. In this case, the clock counts the time since two individuals shared a common mother. Recent studies of dog mitochondrial DNA indicate that domestic dogs are derived from at least five female lines with 95 percent of dogs being derived from three dog "Eves," or first mothers. The overall pattern of the data, based on some 600 dogs representing the full spectrum of

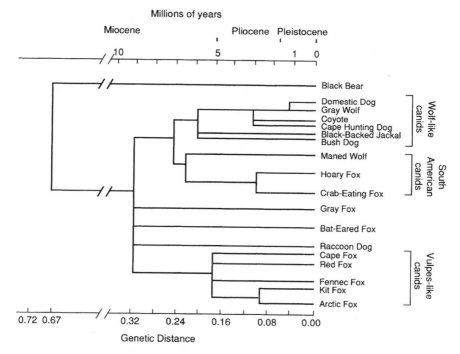

Figure 36. A tree of Canidae based on the degree and pattern of dissimilarities of soluble proteins (or isozymes) found in living animals. From R. K. Wayne et al., Molecular and biochemical evolution of the Carnivora (pp. 465–494), in J. L. Gittleman, ed., Carnivore Behavior, Ecology, and Evolution (Ithaca, N.Y.: Cornell University Press, 1989), based on data from R. K. Wayne and S. J. O'Brien, Allozyme divergence within the Canidae, Systematic Zoology 36 (1987):339–355). Copyright © 1989 by Cornell University. Used by permission of the publisher.

breeds, suggests a domestication of dogs in East Asia (where the genetic diversity of dogs is greatest) between 40,000 and 15,000 years ago. Given the other lines of evidence, the more recent date seems likely.[20]

Similar mitochondrial DNA studies place the dingo in a group with several older dog breeds. These include ancient breeds such as the New Guinea "singing dog," the African basenji, and the greyhound. Also grouped with the dingo by this analysis are a number of more modern breeds.

Clearly, much remains to be learned about the evolution of the dog as a product of domestication. The timing of its initial domestication remains a scientific challenge.[21] However, it is evident that the dog is the first domesticated animal, domesticated by hunting-and-gathering people probably in East Asia sometime in the late Pleistocene—15,000 years ago.

The symbiosis between humans and tamed wolves eventually produced the domesticated dog. This first animal domestication is an excellent example of the process. Humans were in contact with wolves early in the evolution of both species. Early humans (*Homo erectus pekinensis*) and small wolves (*Canis lupus variabilis*) co-occurred between 500,000 and 300,000 years ago in sites in northern China.[22] A cave excavated in southern France contains human-built shelters 125,000 years old, with a wolf skull apparently intentionally placed at each entrance.[23] Humans have had a long history of knowing the wolf as a wild animal, and ample time to experiment with the species as a tamed animal.

Tamed animals are defined as being dependent on people and willing to stay with or close to them. Domesticated animals are raised in captivity and humans control their breeding, territorial organization, and food supply.[24] The domesticated dog is generally thought to have derived from tamed wolves. Initially, perhaps using the methods just described for Australian aboriginal people, wolf pups were tamed and brought into close association with humans.

Remains of tamed wolves look like those of wild wolves. Thus, the evidence for taming is difficult to determine in most cases.[25] J. Clutton-Brock, discussing the evidence for dogs from the Tell of Jericho, notes: "Osteological [bone] characters cannot provide evidence for distinguishing tamed wolf from early domestic dog, and it is in fact still doubtful how clear the distinctions are between wild and tame wolves. But it is of course possible to distinguish wild wolf from fully domesticated dog."[26] Domestication is more a process than an event, but dogs are quite different creatures when compared to wolves—even dogs such as Siberian huskies that superficially resemble wolves. The morphological changes in wolves as they became dogs initially involved foreshortening of the muzzle or nose, crowding of the tooth rows, and a proportional reduction in size of the teeth.[27] One of the earliest examples of foreshortening in wolf skulls was found by in conjunction with a Paleolithic mammoth-hunters' campsite in the Ukraine.[28]

Over antiquity, humans tamed a diverse array of other animals, including hyenas, gazelles, and the foxes tamed by the ancient Egyptians.[29] These and other carnivores such as the cape hunting dog, the coyote, the jackal, and the large cats have been tamed, but never domesticated.

The successful domestication of the dog from the wolf springs from the exceptional fit between humans and wolves with respect to food, behavior, and habitat. Ten thousand to twenty thousand years ago, both humans and wolves in small bands hunted many of the same animals. Both deployed vigorous adults to hunt their prey. Both coordinated their efforts to produce more hunting success. Both were able to drive, weaken, and kill prey that was larger than they were by working in relays to chase and control the flight of the prey. Captures were carried back to young individuals and nursing females. Wolves ate their prey on the spot and regurgitated the meat to their dependents; humans carried it by hand. Both species coordinated their hunting efforts (and their social organization) by a combination of vocal and visual communication. Wolves and humans were organized into a hierarchy of dominant and subordinate individuals. In short, the interactions, organization, mode of communication, and society of a small human settlement would not have been totally alien to the wolf.[30]

The domestication of tamed wolves did not have to overcome as much of the intrinsic nature of the animal as the domestication of some other animal might have. Several conditions are necessary for an animal to be domesticated. First, the young animal has to be physically tough enough to survive the trauma of removal from its own mother (probably before it is weaned). Further, it must be hardy enough to adapt to a new diet and environment. Second, the species behavior must be comparable to human social structure. What is needed is a social animal with dominance hierarchies that will accept humans as leaders and that will remain imprinted on humans for life. Wolves conform well to these conditions, whereas most other animals do not.

Practical considerations enter into the picture as well. The animals should not bolt into instant flight (as antelopes and deer do). They should be amenable to close contact, and to being penned or herded together. They should be useful to humans in some way. For many captive animals, the basic factor is their role as an easily maintained, nonspoiling source of food that can supply meat when required. Their breeding should not require complex or hard-to-create conditions.

Particularly for herded animals, they should be reasonably placid, versatile in their feeding habits, and gregarious so that individuals will stay together to be herded.

The husbandry of domesticated animals generates modification of landscapes to accommodate creatures with this relatively unusual set of attributes, eventually at the expense of other animals that do not have the domestication attributes in their niche description.[31]

Our portfolio of domesticated animals is very small compared to the overall diversity of animals. The length of the list varies to some degree depending on how rigorously one defines "domesticated." Generally, domesticated animals have four principal attributes: humans control their breeding; they provide a useful service or product to people; they are tame; they have been genetically selected away from the attributes of their wild counterparts.[32]

Goats and sheep were the next animal domesticates after dogs. Like humans and wolves, goats and sheep interact in dominance hierarchies, making the animals more likely to conform to human society. Hunters controlling prey herds and eventually becoming nomadic pastoralists for tamed animals might have domesticated goats and sheep independently of plant domestication. Alternatively, the sedentary life of the farmer could have motivated the taming and domestication of sheep and goats.

The Polynesians traveled the Pacific with a portable "kit" of domesticated plants and animals that they used to enhance their chances of survival. The equivalent kit for the European colonizers of different parts of the world was mostly developed in the region that is now Iran and Iraq, beginning about 10,000 years ago.

Many of the animals in the European kit were developed in a zone of overlap in the vicinity of what is now termed the Near East or southwestern Asia. This area was one of the centers of plant domestication as well. The flora contained an unusual number of large-seeded plants that would make profitable the initial attempts to domesticate them. The region is also in an overlap zone in the natural ranges of a set of animals that eventually became the mainstays of the European domesticate portfolio (Figure 37). These species are the goat, sheep, pig, and auroch (the species from which domestic cattle were derived). Studies on the mitochondrial DNA of these domesticated creatures indicate two regions of early animal domestication, one in the "Fertile Cres-

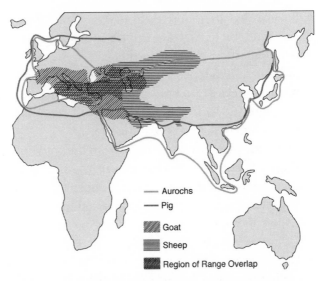

Figure 37. Probable distribution of the progenitors of the four principal species of livestock (auroch, pig, goat, and sheep) at the end of the Pleistocene. The widespread ranges of the species overlap in western Asia. From J. Clutton-Brock, A Natural History of Domesticated Mammals *(Austin: University of Texas Press, 1989); after E. Isaac,* Geography of Domestication *(Englewood Cliffs, N.J.: Prentice-Hall, 1970).*

cent" region and another farther east in Asia.[33] Along with the dog, the following are some of the important animals in the European package of animal domesticates:

- Goats (*Capra hircus*) from wild bezoar goats (*C. aegagrus*): Goat bones and pieces are difficult to distinguish from those of sheep. Goats appear to have been domesticated at about the same time as sheep in western Asia.[34] In early sites where both occur, goats seem to have been a greater supplier of meat than sheep were.[35] Mitochondrial DNA evidence indicates that modern goats are possibly the result of multiple domestication events following an initial such event about 11,000 years ago. From their genetics, it appears that goats have been broadly transported more than other domesticated animals, at least when compared to cattle—invoking images of early goat-herding pastoralists migrating over wide areas.[36]

- Sheep (*Ovis aries*) derived from Asiatic mouflons (*O. orientalis*): Sheep originated in western Asia and Greece, possibly with a second domestication event farther east.[37] Remains of sheep from Iraq indicate that the species may have been tamed as early as 11,000 B.P. The species may have also descended in part from European mouflons (*O. musimon*), if the European mouflon is not a feral relic population of the first domestic sheep.[38]
- Domestic pigs (*Sus domesticus*) from the wild boar (*Sus scrofa*): Pigs were domesticated in western Asia and southeastern Europe slightly after the domestication of sheep (10,000 years ago). Perhaps from an independent domestication event, pigs were also in Asia at about the same time.[39] They may possibly have been in New Guinea as early as 9000 B.P.[40]
- Domestic cattle (*Bos taurus*) from aurochsen (*B. primigenius*): The earliest record of domesticated cattle is from Turkey around 8,400 years ago. Domestic humped cattle such as brahmins (*B. indicus*) are derived from an Indian subspecies of the aurochsen.[41]
- Horses (*Equus caballus*) from *Equus ferus:* The horse appears to have been domesticated by the end of the Neolithic period, perhaps around 5000 B.P. The mitochondrial DNA from horses indicates a large number of "mothers," implying multiple domestication events.[42] Because the species was domesticated later in human history than other animals, it appears that the technology of domesticating horses spread across Eurasia. In older domesticates (notably goats), the domesticated animals themselves were spread by human activities.[43] Initially, the horse was domesticated as a meat animal. Its first use as a riding animal was in southern Russia and central Europe. The related domesticate, the ass (*Equus asinus* derived from *E. africanus*), was being bred in captivity in Egypt by 6000 B.P.[44]

The dog was developed earlier in human prehistory than the rest of the European kit. Among this animal's array of potential uses to humans (part of a hunting team, nonperishable food item, boon companion, self-heating blanket, camp clean-up crew), the dog's ability to herd and control grazing animals would make it a considerable aid in the taming and domestication of the large herbivores that were the next domesticated animals.

Goats and sheep are two species that humans, particularly those who had the domesticated dog as a partner, could have easily tamed and domesticated. The process of domestication would have developed from hunting goats and sheep, to controlling the movement of goat and sheep herds and thus keeping them readily available, to working as nomadic pastoralists moving herds from one grazing area to another.

Domestication of different animals can be seen as a logical historical progression.[45] Species domesticated in the preagricultural phase of human society include the dog and potentially three species subsequently herded with the help of dogs: the sheep, the goat, and the reindeer (*Rangifer tarandus*). Domestic reindeer are not strongly modified physically from the wild type, and the timing of their domestication is ambiguous. People in the far north of Asia who own domesticated dogs have been eating reindeer for a long time. When hunting reindeer became herding tamed reindeer, followed by herding domesticated reindeer (as well as using them as draft animals to pull sleds) is difficult to determine.

These preagricultural domesticates were followed by species domesticated in the early agricultural phase, including cattle and pigs (in the "standard" Western portfolio), along with water buffalo (*Bubalus bubalus*), yaks (*Bos mutus*), and benteng (*B. javanicus*) from Asian locations. These large, potentially dangerous animals are difficult, particularly the wild species. They probably came into close contact with people as crop marauders.[46] Cattle may have been tamed by acclimating them to the presence of humans by providing salt, or simply by being a nonthreatening presence among them. Some authorities feel that the early cattle were initially tamed for use in religious ceremonies or as animal sacrifices.

Finally, there is a collection of species domesticated for transport and labor. Domesticated cattle were early draft animals. The horse initially was used as a meat supplier, then as a superior draft animal relative to cattle, and then as a riding animal. This last capability allowed mobility for military conquests and expansion by the horses' masters. Other species in this category are elephants (*Elephas maximus*), camels (*Camelus dromedarius* and *C. bactrianus*), asses, and onagers (*Equus onager*).[47]

An additional category that is somewhat removed from this temporal ordering is what are called the pest destroyers.[48] These would have

adapted to humans from a symbiotic relationship where the domesticated animals were predators on the pests that ate stored grain or fed
on other products of human society. Notable in this group is the house
cat (*Felis domesticus*). Other examples are the domestic ferret (*Mustela
putorius*) and the Egyptian mongoose (*Herpestes ichneumon*).

Certain archeological sites have provided remarkable chronologies
of the domestication of crops and animals. Several of these sites in the
Middle East are called tells—stratified hills made by the inhabitants of
ancient city buildings on the ruins of the previous buildings. In a tell,
layers build up as the mud bricks used to build houses and other buildings break down over time; the town location grows in elevation on its
own mud and ruble. Mud from the mud bricks is not recycled as are
the stones used in construction in some other areas. Older tells have
considerable height.

One important such site is the Tell of Jericho, which provides a long
chronology from the time of the earliest domestication of plants and
animals.[49] The cultural period in this region, referred to as the Natufian period, saw the beginning of the demise of the hunter-gatherer
people who lived near the ancient biblical city of Jericho. After the
Natufian, at around 10,000 B.P., Jericho was a town of about 2,000 people with stone-walled outer defenses and mud-brick houses. The inhabitants of Jericho at this time did not make pottery (hence the name
of the period, Pre-Pottery Neolithic A). They were farmers of two early
cereals (emmer wheat, *Triticum dicoccum,* and hulled two-row barley,
Hordeum distischum).[50] They appear to have been hunters as well and
left behind a diversity of animal bones in the deposits at Jericho. Their
principal food seems to have been gazelles, along with a substantial
number of foxes.[51]

Over the next thousand years, the climate in the region became
more arid. The settlements in the area surrounding Jericho tended to
concentrate around water supplies. At Jericho, new culture developed
that is now called the Pre-Pottery Neolithic B culture. This culture did
not have pottery-making technology either, but it did have domesticated goats and sheep. Bones of these animals evidence a shift in diet
(Figure 38): the hunting Pre-Pottery Neolithic A people ate gazelles
and foxes; the Neolithic B people ate mostly sheep and goats.[52]

Thus, the deposits from the Tell of Jericho would support an interpretation of early-domesticated grains being followed by early-domes-

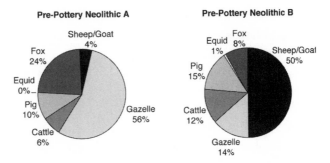

Figure 38. Percentage of animal remains from the Pre-Pottery Neolithic A and Pre-Pottery Neolithic B cultural periods at the Tell of Jericho. The percentages are based on the total absolute numbers of bones and teeth found in each layer. From Pre-pottery Neolithic A, 531 such bones and teeth were identified; from Pre-pottery Neolithic B, 773. The transition of the occupants was from an apparently hunting and farming people who used gazelles and foxes for meat to humans with domesticated goats and sheep as meat. "Equid" is a collective term for members of the genus Equus *and includes the horse, ass, and onager. From data in J. Clutton-Brock, Carnivore remains from evacuations of the Jericho Tell (pp. 337–345), in P. Ucko and G. W. Dimbleby (eds.),* The Domestication of Plants and Animals *(Chicago: Aldine, 1969).*

ticated sheep and goats. Still, the use of both plant and animal domesticates in Jericho, while it was very early in human agricultural history, may have developed from already-domesticated species brought in from elsewhere in the region. Goat herding or sheep herding, perhaps with the help of dogs, as an endeavor independent from plant-based agriculture, could have developed as nomadic animal herding. This practice is likely to leave sparser evidence than the remarkable Tell of Jericho and the other tells in the region.

To prosper, the package of domesticated plants and animals that we have currently developed demands considerable alteration of most natural landscapes. However, domesticated animals can serve as active agents to produce some of this necessary change. Goats are browsing animals that feed on leaves, twigs, and branches of shrubs and trees. They can survive and modify the vegetation found in relatively dry environments. Sheep graze on grass and can maintain pastures created in what would otherwise be forested environments. Cattle feeding in forests create a graze line to the height of their heads by eating everything they can reach. Grazing pigs in forests remove ground vegetation, seeds, and tubers and stir the soil. Horses as draft animals can pull plows through the soil. By judicious movement of grazing animals onto landscapes (perhaps along with setting wildfires) at different times and in different sequences, people, even without modern machines or even metal implements, can radically alter landscapes.

The ancient city of Petra, the remarkable archeological site used as a set in the 1989 movie "Indiana Jones and the Last Crusade," contains hyrax middens that reveal the sequence of changes associated with the use of domesticated plants and animals in that Jordan valley.[53] Hyraxes (*Procavia capensis*) are mammals about the size of a large rabbit. Their closest living relatives, albeit quite remote, are elephants. Like the packrats discussed in Chapter 4, hyraxes build middens that preserve plant material and pollen grains. In the region of Petra, initial inspections of hyrax middens indicate a degradation of the surrounding landscapes from a Mediterranean forest with oak (*Quereus*), pistachio (*Pistacia*), olive (*Olea*), pine (*Pinus*), and juniper (*Juniperus*) to a "maquis" (a degraded forest) and then to a "garigue" (an even further degraded shrubby forest) by the second century A.D. After the collapse of Byzantine Petra, intensified grazing by livestock eventually reduced this vegetation to Mediterranean "batha" (very open shrubby vegetation), and in some cases to sparse grassland.

The species domesticated in the Near East were drawn from wild progenitors that ranged over much of Eurasia and northern Africa and were able to prosper in a wide range of environmental conditions. As domesticated plants and animals become more and more important to a society, the task of meeting the biological needs of these organisms has increased their constraints on human activities.

The heads of cereal grains such as barley and wheat have relatively few seeds. Availability of enough seed grain to plant the next year requires a relatively large seed store over the nongrowing season, a time when food supplies of a cereal-grain dependent people diminish. The use of cereal grains as a mainstay requires considerable social discipline to ensure the availability of sufficient seed to be planted at the beginning of each growing season. The lack of such discipline can be calamitous.

The sudden effect on a human society of access to domesticated animals is illustrated by the progression of changes in Cheyenne society. European settlers had reintroduced the horse to North America and Cheyenne culture thereby became acquainted with the domesticated horse. The Cheyenne, a forest people, migrated onto the Great Plains of the central United States, in response to perceived opportunity for a people with horses. They were also responding to pressures from other Indian tribes, who in turn were being pushed westward by European settlement of the eastern part of the country.

The Cheyenne became nomadic pastoralists-hunters. They hunted the great bison herds of the plains from horseback and used their horses to accompany these migratory herds. Becoming dependent on their horses, the Cheyenne were forced to adjust their culture to the distinctive patterns of behavior of their animals.[54] Elliot West has summarized their situation as follows: "We are used to thinking of the plains Indians as nomadic hunters who patterned their lives to follow the herds of bison that supplied so many of the Native Americans' own immediate needs. It is just as accurate to imagine them as arranging their movements according to what their horses had to have. They spent the year chasing grass, basically."[55] From residing permanently in relatively large settlements, the Cheyenne adopted a nomadic life, living in small hunting bands with large numbers of horses (from five to as many as thirteen per person).[56] They would meet annually to maintain parts of their former settlement rituals and social interactions. They were in the process of setting themselves up as middlemen, exchanging goods over long distances via their horses, when Europeans began their occupation of the Great Plains. There is no question that the Cheyenne were substantially changed by the horse, by both its needs and its advantages.

For the early grain farmers and the Cheyenne nomadic hunters, acquiring domesticated plants and animals transformed the way they supported their populations. Simultaneously, their dependence on domesticated species implied changes in human culture to accommodate these dependencies. The reciprocal interactions between human cultures and domesticated species also greatly modified the landscapes on which they lived.

Domestication was a Stone Age "biotechnology" that drove our species to alter the space scale of our planet. The story of the dog and of species domestication is about the landscape change wrought by humans through all of our written history and earlier, and it foretells possible future changes. The husbandry of animals and the stewardship of the land require a great deal of knowledge of animals, plants, and landscape processes. Witness the continuity from the initial understanding of how to use well the repertoire of domesticates developed by societies over millennia to the science-based curricula of modern agricultural universities. Even short-term errors in land management can produce long-term losses in environmental conditions.

Like the Polynesian portable biota, the European domesticated biotic kit can greatly modify landscapes when applied to new territory. For a classic example, consider the extensive areas of the United States developed through government programs to bring agricultural production to the western frontier of the new nation. The Homestead Act of 1862 provided land to settlers willing to build on new farms. Settlers were allotted 160 acres (~62 ha) of land, which became theirs if they farmed it for five years. The change in Wisconsin, which was settled by homesteaders, has been the focus of several studies of landscape fragmentation (Figure 39).

Small woodlots, with a large portion of their area in forest edge, have a high representation of sun-loving plants that prosper in the well-illuminated edge and adjacent area. The general term for this is edge effect—the often-unique assemblages of plants and animals found along the abrupt transitions from vegetation dominated by one life form to vegetation dominated by another (such as the boundary between field and forest).[57] Edges often have significantly different microclimates, particularly with respect to light levels, wind velocity, and moisture conditions.

A large intact landscape of forest and an equivalent area of forested woodlots differ in the amount of edge per unit of forest area. They also differ in the diversity and abundance of species. Based on a detailed computer model of expected vegetation in Wisconsin woodlots of various sizes, different trees use different parts of the woodlots (Figure 40).[58] Some species, such as American beech (*Fagus grandifolia*), prosper in the cooler, moister, and shadier centers of the woodlots (Figure 40A). The eastern edge of a plot receives morning sun at a cooler time, after the trees have restored their water balance overnight. The western edges receive the same amount of sunlight, but it comes in the hot

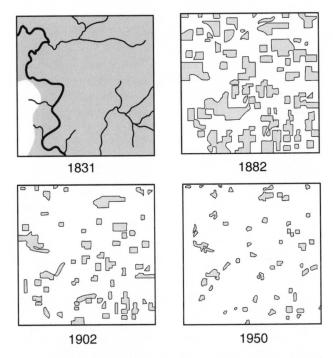

Figure 39. Reduction and fragmentation of Cadiz township, Green County, Wisconsin, from 1831 to 1950. The area of a township is 6 × 6 miles (about 10 × 10 km). There would be four 160-acre homesteads per square mile. Because the homesteads were laid out "on the square," regularity in orientation of the edges is evident in any of the years. From J. T. Curtis, The Vegetation of Wisconsin: An Ordination of Plant Communities (Madison: University of Wisconsin Press, 1959).

afternoon, causing the plants to use more water to photosynthesize and grow. Ash (*Fraxinus americana*) grows better on the eastern woodlot edges (Figure 40B), while basswood (*Tilia americana*) can tolerate the effectively hotter and drier conditions on the western edges (Figure 40C). A large woodlot would have more American beech than a number of small woodlots that added up to the same area. The collection of small woodlots would have more edge, and hence more basswood and ash.

These edge-and-area effects have been observed often, and in enough studies so that they are expected for landscapes everywhere. Thus, the proportions of habitats over a given area of fragmented land-

A. American Beech

B. Ash

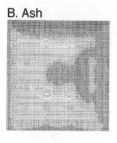

C. Basswood

Figure 40. Distribution of trees in a Wisconsin woodlot as simulated by the FOREST model. The darker the shading, the more abundant the tree. Distribution of A, American beech (Fagus grandifolia); B, ash (Fraxinus americana); C, basswood (Tilia heterophila). The top of the page is north. From J. W. Ranney, Edges of forest islands: Structure, composition, and importance to regional forest dynamics, doctoral dissertation (Knoxville: University of Tennessee, 1978).

scape change according to landscape geometry. Depending on the fragmentation and spatial patterns used to develop a park or preserve, equal areas of a landscape can have very different utility with respect to maintaining a particular species of plant or animal. Therefore, the geometric patterns of landscapes influence priorities for the acquisition of land as well as for the design of nature preserves and parks intended to conserve biotic diversity.

Fragmented landscapes can be thought of as patches or islands of suitable habitat in a matrix of unsuitable, even inhospitable, habitat. They have been likened to islands to develop a general theory of the expected effects of fragmentation on animal populations. While this is an appealing concept and one that has merit in focusing on the nature of fragmented landscapes, an island archipelago is quite different from an assemblage of small habitat patches. For one thing, the time scales of the processes that are involved are different.

In most cases studied, the islands have been in existence for thousands of years (for example, many near islands, isolated by the rising of the seas since the last ice age) or millions of years (many distant islands, such as the Hawaiian Islands). Fragmented landscapes do occur naturally, but most are the result of relatively recent land clearing to support agriculture.[59] Even for mobile species, arrival at far oceanic islands occurs infrequently—once in a millennium or even less often. Arrival rates for such species to fragmented agricultural landscapes are more frequent—yearly and often much more often.

Two critical issues associated with the fragmentation of landscapes arise for conservation ecologists. First, what is the difference between a number of fragments of habitats and an equivalent area of unfragmented habitat? Second, how large does a tract of suitable habitat need to be to support a sustainable population of a given species?

The effects of fragmentation vary with the life-history attributes of the species. Clearly, the habitat of the summation of forest fragments would differ from an equivalent area of intact forest. The former would have a greater proportion of edge-like conditions. Even without habitat effects, the size of the patches has an effect on the success of animal populations. From our stories of the ivory-billed woodpecker, Bachman's warbler, and other examples, it is obvious that some species require remarkably large areas for the underlying ecological processes to provide them with habitat.

Of course, successful breeding in a small population does not guarantee a sustainable population.[60] Small populations become inbred over time and lose their genetic variability and overall viability. Wildlife managers of game preserves in Africa (and elsewhere) work to overcome the inbreeding problems for species such as the white rhinoceros (*Ceratotherium simum*) by moving them from location to location. Zoos have similar programs for a number of the rare and endangered species that they maintain. It is not clear what number should constitute a minimal breeding unit, although a breeding population of at least fifty individuals often is used as a rule of thumb.

Owing to agricultural land conversion, fragmented landscapes—with small patches of habitat, with potentially isolated populations, and with regular losses of the species that require larger areas—are becoming the norm in many parts of the world. Some species will thereby be disadvantaged more than others; the loss is not random and it eliminates a segment of the biota. The tendency is to lose first the animals that require large areas: such striking creatures as tigers, jaguars, and other big cats. The example of the Leadbeater's possum reminds us that less striking and more secretive small creatures also are selectively lost from increasingly fragmented land situations.

Weedy annual plants have matched their life cycles to their host crops so that their seeds are harvested when the grain is planted in the next year's seeding. Poisonous or inedible plants prosper as grazing animals remove their competitors from pastures and grasslands. Insects that are able to feed on the crops prosper. Studies of the attributes of weeds and pests form a large and important body of scientific work.

The biological issues involved with how a species becomes a successful weed or pest have contributed significantly to our understanding of the process of evolution.

Landscape changes that accompany the successful spread of domesticated plants and animals strongly affect native flora and fauna. Organisms that are adapted to the new environment prosper and become weeds, pests, and parasites. Less-adapted species decline in numbers and perhaps even become extinct.

Modern society depends for its sustenance almost totally on a relatively small number of species of plants and animals. Starting with the dog, the animals that humans have domesticated have provided the capability to alter the face of the landscape to a remarkable degree. Deployment of the power provided by these creatures to alter the land and to help in the cultivation of crops has been premeditated. The care and promotion of domesticates has undoubtedly shaped the nature of civilization. It remains to be seen how wisely humankind can use the capabilities provided by animal husbandry and agriculture.

The alteration of natural vegetation and its replacement with agricultural lands is a process that has been ongoing ever since the invention of agriculture some 10,000 years ago. For the hunters and gatherers who used landscape fires as a management tool, the time of substantial landscape alteration was perhaps much earlier. Today's technological societies have the potential to convert land rapidly and also to harvest landscapes. Scientific studies indicate that the large-scale consequences of these actions may be significant. That the ultimate results of our actions may be irreversible speaks to a need for far better understanding of the consequences of the changes we are making to the surface of our planet.

10 *The Gentle Invader*

There is also a species of hare, in Spain, which is called the rabbit;
it is extremely prolific and produces famine in the Balearic islands
by destroying the harvests . . . It is a well known fact that the
inhabitants of the Balearic islands begged of the late Emperor
Augustus the aid of a number of soldiers, to prevent the rapid
increase of these animals.—Pliny the Elder, A.D. 81

At the dawn of recorded history, the distribution of the European
rabbit (*Oryctolagus cuniculus*) in Europe was confined to what is now
the Iberian Peninsula. In 1100 B.C., Phoenicians sailing to the peninsula found large populations of these rabbits, a species they had not
previously encountered. They named the country after the rabbits;
translated, that name became Hispania, or Spain.[1]

Populations of rabbits existed too in northwestern Africa, but inasmuch as they do not show up in the fossil deposits until Neolithic
times, humans may have transported them there.[2] If so, this prehistoric
human introduction of the rabbit to Africa was a harbinger of things to
come. Through its interaction with humans, the European rabbit has
now spread to every continent except Antarctica, as well as to hundreds of remote islands.[3] The outbreak in the Balearic Islands noted by
Pliny in the first century was also an indication of the future, for the
rabbit's reproductive potential often allows introduced populations to
explode to plague proportions.

The European rabbit (Figure 41) is a lagomorph, one of a group of
small mammals that differ by several anatomic features from the ro-

dents they resemble.[4] One ecologically significant attribute of lagomorphs is the habit of coprophagy, the redigestion of feces—an adaptation for feeding on grasses and other coarse vegetable material.
Other grass feeders, such as cows, have chambered stomachs: plant
material is initially fermented in one of the chambers, then chewed a
second time and moved to another stomach chamber for further digestion. In rabbits and other lagomorphs, with the first passage through
the animal's digestive system fecal material is produced as moist pellets
that are eaten to pass through the gut a second time. The second passage produces a second batch of more fibrous pellets that are discarded.
The coprophagous habit is essentially an external multistep digestion
process analogous to that internalized in ungulates. Coprophagy makes
the rabbit an efficient converter of vegetable food into rabbit flesh.
Nonetheless, the rabbit does rely on more succulent and less fibrous
food than most ungulates.

The rabbit is also portable, tasty, and docile; it reproduces so prolif-

*Figure 41. European rabbits (Oryctolagus cuniculus) at a water hole near
Wardang in Australia during a drought in 1938. Photograph by Bill Mule,
provided by CSIRO Wildlife and Ecology Division, Australia. Reproduced by
permission of CSIRO Australia © CSIRO.*

ically in captivity that some farmers in nineteenth-century England felt that both sexes could bear young.[5] Thus, rabbits are obvious targets for domestication by humans.

The European rabbit differs from other rabbits and hares in its unique habit of digging extensive communal burrows.[6] The species name, *cuniculus,* in the rabbit's scientific binomial is from the Latin word that refers to such burrows as well as to the animals that make them. A consequence of its colonial habit is that rabbits prosper in enclosures. However, their burrowing habit makes it difficult to completely contain them in these same enclosures. Rabbit colonies are prone to produce escapees that spread the species across a region.

European rabbits and their lagomorph relatives, hares, were popular meats for the Romans, who raised rabbits in large pens now called warrens—some apparently over a half-mile in diameter.[7] The rearing of rabbits in warrens had the advantage that the species could be bred on land that was otherwise useless, an advantage that continued through their history as a tamed and eventually domesticated animal.

The Romans did not actually produce domesticated breeds of rabbits. They fed and fattened captured animals, but largely allowed the species to live as in the wild (except, of course, for containing them in enclosed hutches and warrens). It is even possible that the regular capture of rabbits within a large warren would accidentally "undomesticate" the species; the tamer rabbits would be caught, leaving the more wary individuals to reproduce.[8] The Romans spread the European rabbit from Spain throughout the Mediterranean. It was introduced into Italy at least by A.D. 230.[9]

Medieval monks in French monasteries developed domestic breeds between A.D. 500 and A.D. 1000. Along with finding adult rabbits to be excellent food, the monks had the additional incentive that unborn and newborn rabbits (called *laurices*) were considered aquatic creatures. They were defined as fish, not meat, and could be eaten during the fasting period of Lent.[10] The domestic rabbit is considerably larger than its wild Spanish ancestors, and it is generally assumed that the initial selection in domesticated rabbits was for size. There is no mention of colored varieties of rabbits until the mid-sixteenth century.[11]

As noted in the previous chapter, Polynesians and earlier prehistoric sailors of the South Pacific often transported dogs for food. Similarly, throughout the association of the European rabbit and humans, sailors have transported rabbits. They also have released rabbits on islands to provide provisioning for subsequent voyages. This practice started with early Phoenicians and was continued by Roman sailors who brought rabbits to the Balearic Islands around 30 B.C. and to Corsica by A.D. 230.[12] Over history, subsequent sailors have spread rabbits to over 800 islands, worldwide.[13]

The European rabbit was not found in the British Isles until after the Norman invasion of 1066. Rabbit warrens were recorded frequently on islands off the coast of England in the late twelfth century, and on mainland England by the middle of the thirteenth century.[14] By 1399 rabbit appeared on the menu for the coronation of Henry VI, and it was prominently listed at other great feasts (for example, the installation of the Archbishop of Canterbury in 1443 and of the Archbishop of York in 1465).[15]

The features that made the rabbit a candidate for domestication would eventually make it a superior invader of new territory—particularly of landscapes that had been altered by human activities. In addition to the transport by sailors to remote islands, the species has been introduced successfully to agricultural or pastoral landscapes in various parts of the world.

The origin of the rabbit on the Australian continent is an excellent example of the invasion pattern of a successful exotic animal. European rabbits and European people arrived to become residents of Australia in January 1788, with the First Fleet at Botany Bay.[16] Additional members of the species were brought to the colony with subsequent human arrivals. In 1827 Peter Cunningham, a naval surgeon, wrote, "Rabbits are bred about houses, but we have no wild ones in enclosures, although there is a good scope of sandy country on the sea coast between Port Jackson and Botany Bay fit for little else than goat pasture and rabbit warrens."[17]

Subsequently, the rabbit was shipped to several Australian ports. It was soon a fixture on various properties and was introduced to the island of Tasmania (in 1820), where it eventually became established as a feral animal.[18] At some point in the process of escaping and becoming a feral species in Tasmania, it changed its coloration. The initial rabbit introduced into the Tasmanian colony was silver-grey (a variety named for its coloration). By 1869, owing to additional introductions, it had "the rusty grey colour of the wild rabbit in England."[19]

On Christmas night 1859, Thomas Austin in Melbourne received the twenty-four rabbits (along with sixty-six partridges and four hares) that his brother, James, had loaded into the Black Ball clipper *Lightning* in Liverpool.[20] From his estate called Barwon Park, Winchelsea, in Victoria (near Geelong), Thomas may have released some of his rabbits immediately. This would date the first successful release of rabbits into the wild on mainland Australia. Certainly by 1862, Thomas Austin's rabbits, held in a fenced enclosure, were thriving and soon became so numerous that they began to chew the bark from the trees. At some point they either escaped or were released. Rabbits were damaging neighboring farms by the mid-1860s and spread rapidly through the district. This successful introduction inspired imitation among the hunters of the landed gentry in nearby areas. Rabbit hunts and rabbit coursing became a popular field sport among the wealthy in several locales where rabbits had been loosed on the land.

Without native rabbit diseases, the animal moved rapidly into a climatic and habitat context that matched its requirements very well. The population of wild rabbits increased explosively in the decade of the 1860s (Figure 42) and moved as a wave over Australia. The rate of advance over rangelands was as much as 60 miles (100 km) per year! This is the most rapid spread known for any colonizing mammal anywhere on Earth.[21] The 1860s in Australia was a decade of drought. Sheep populations crashed to as much as half their former density and pastures were degraded. That the rabbit population increased so rapidly in this period testifies to the importance of landscape condition to the success and spread of an introduced species.

By the late 1860s, rabbits had become so prevalent that they were considered serious pests. One landholder, William Robertson, paid trappers and professional rabbit killers £9,500 to eliminate well over two million rabbits in just one year on his property. To stop the spread of the rabbit, a series of "rabbit-proof" fences were built across the incredible distances that make up interior Australia. The lengths of these fences are staggering. One fence across New South Wales was 700 miles (1,100 km) long; the so-called No. 1 fence was 1,100 miles (1,822 km); and the No. 2 fence was another 700 miles (1,100 km). Between 1888 and 1890, the entire New South Wales–South Australian border was fenced, as was the Queensland–South Australian border.

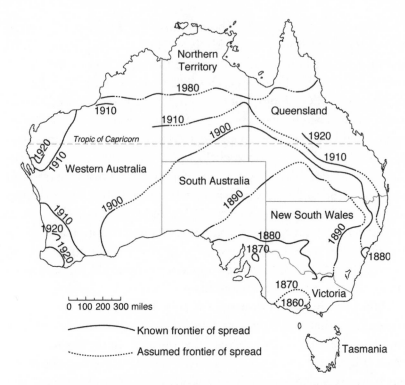

Figure 42. Spread of the European rabbit in Australia. From K. Williams et al.,
Managing Vertebrate Pests: Rabbits *(Canberra: Australian Government
Printing Office, 1995); originally published in E. Stodart and I. Parer,*
Colonisation of Australia by the Rabbit *(Canberra: CSIRO Division of
Wildlife and Ecology, Australian Government Printing Office, 1988).*

Rabbit fences were expensive. They were constructed using 1.25-inch (32-mm) mesh wire 40 inches (1 m) high and 6 inches (15 cm) underground. Despite their great cost, they were largely ineffectual. Often the rabbits advanced beyond the fences before they could be constructed. Fences of these lengths are difficult to maintain across the vast never-never of Australia; breaks occurred regularly. Other control attempts, notably poisoning and trapping rabbits to reduce their numbers, also poisoned and trapped Australia's population of native marsupial mammals.

By the 1930s rabbits had largely reached their present range, occupying the areas that are considered suitable habitats for the species. Today in Australia, rabbits occur mostly south of the Tropic of Capricorn on well-drained soils. They are not generally found in dense forests or at higher elevations.[22] Thomas Austin's release of European rabbits to the Australian continent, which may have seemed a good idea at the time, became an ecological disaster.

Why was the European rabbit such a successful invader? Why was it able to establish a vigorous population on a continent stocked with locally adapted mammals that were potential predators and competitors? The European rabbit has several features that predisposed it to conquer new territory in Australia. It has a high reproductive rate. Its habit of building communal dens provides protection from predators. More important, Australia was filling with European domesticated sheep and cattle and was being converted from its natural state into prime rabbit habitat.

Successful invaders often have high reproductive rates and the ability to use habitats that have been altered by novel disturbances, such as human land conversion. If they arrive at new areas where they are not exposed to disease, parasites, or predators, the invaders can increase explosively. We see these aspects best in the case of distant islands.

If we turn now to New Zealand, one of the possible factors in the disappearance of the moa (Chapter 8) and other species was the introduction of the kiore, or Polynesian rat. A great number of species have been lost from New Zealand, several of which, including the moa, were remarkably different from any other creatures on Earth. At the same time, invasive species such as the kiore have come to the islands.

What is the present balance of the changes in diversity of New Zealand species? The country has been populated by a large number of introduced exotic species. Of its inland fish, 27 are native species and 30 are exotic.[23] Of its birds, 155 are native and 36 are exotic.[24] New Zealand today has about 2,500 native species of plants and has gained some 1,600 new exotic species.[25] Invasion has served to increase species diversity there. Indeed, it has gained species unlike any of those that were previously present. New Zealand has exchanged a richness of endemic species for an even greater richness of native plus introduced species. This "gain" in biotic diversity has been the world's loss. While the kiore and virtually all the other introduced species are found elsewhere, the moa and most of the other species lost were found nowhere but New Zealand.

The islands of the South Pacific (and elsewhere) have been greatly altered by introductions in the past thousand years or so. The voyaging Polynesians brought their hunting activities and agricultural practices, and their novel and effective predators: the dog, the rat, and the pig.[26] The rate of species invasion accelerated with European settlement and agriculture, and increased exchanges of exotic species. At the dawn of the twentieth century, commerce among the islands, and visitations by ships to pick up copra (dried coconut) and other products, spread insect crop pests from Asia and nearby islands to the more distant islands of the South Pacific.

In the pre-DDT era, several of the South Pacific islands initiated biological control of pests via the deliberate introduction of predators and parasites. The success of some of these introductions was remarkable.[27] For example, in the early 1920s a small purplish moth (Figure 43A) began to spread across the kingdom of Fiji. It devastated coconut plantations and endangered the copra trade, the second most important industry in the country. Coconut palm plantations attacked by the caterpillars of this moth looked as if they had been swept by wildfire. The little purple coconut moth (*Levuana iridescens*) appeared to have no natural enemies, and with increased contacts in the archipelago it was spreading more or less unchecked through Fijian plantations. *Levuana* was formerly only found on one of the Fijian islands, Viti Levu, where it had limited the success of coconut production.

While the presence of the moth was considered to be the product of a curse of an old chief, it was first recorded in the 1860s, contemporaneous with the arrival of the sandalwood traders.[28] Sandalwood produced something of a gold rush in the South Pacific, with small freighters making considerable money cleaning out the sandalwood from Fiji and other islands. Initially, it was not clear if the moth was native to Fiji or was brought there as an unwitting consequence of the new commerce.

Hubert W. Simmonds, who was to work in Fiji on insect pest problems for forty-five years, spent the early part of his career trying to locate the origin of *Levuana*.[29] His hope was to find a predator or parasite that could control the animal. Simmonds spent most of 1922 sailing the islands in some of the more remote parts of Fiji, searching for the insect and its possible parasites. His lack of success in finding the pest in distant Fijian islands indicated that the species had been introduced from elsewhere. From 1923 into 1925 Simmonds searched for the species on islands to the west of Fiji (New Guinea, the Bismarcks, the Solomon Islands, and the New Hebrides)—to no avail.

A relatively similar moth (Figure 43B), which was a major pest of coconut in Malaysia and Indonesia, had a parasitic wasp (Figure 43C) that laid its eggs on caterpillars. Simmonds theorized that possibly this parasite could control *Levuana*. He and his wife spent from February 1925 until January of the next year searching for the moth in Java and Malaysia. The problem was that the Malaysian moth populations exploded to wipe out plantations, but most of the time the species was rare and the parasitic wasp was even rarer. He finally found an outbreak in Kuala Lumpur, Malaysia, and shipped large kerosene tins with small potted palms, moths, and wasps, back to Fiji. He was searching Java for another moth outbreak when he was summoned home: the first wasps he had shipped appeared to be successful and

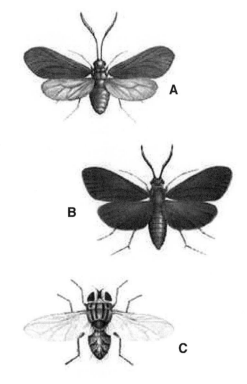

Figure 43. A,The small purple coconut moth (Levuana iridescens); B, The Malaysian coconut moth (Artona catoxantha); C, The wasp parasite (Ptychomyia remota) of the Malaysian coconut moth, found near Kuala Lumpur and shipped to Fiji to control the Levuana. From H. W. Simmonds, My Weapons Had Wings: The Adventures of a Government Entomologist Based in Fiji for Forty-Five Years (Auckland: Percy Salman, Wills and Grainger, 1964). Courtesy of the Fiji Society.

were wiping out *Levuana*. At the time of Simmonds' retirement in 1937, *Levuana* was a rare species in Fiji. He never did determine where the creature originated.

Control of the little purple coconut moth was a success story. Yet some of the ad hoc attempts to apply biological control were disastrous. The mongoose, immortalized as a cobra killer in Kipling's "Rikki

Tikki Tavi," was spread throughout the colonial holdings of the tropical British Empire and elsewhere as a rat control agent—to the great disadvantage and subsequent extinction of several native island animals. Similar nightmare cases involved such creatures as the cane toad (*Bufo marinus*), a poisonous, predatory toad that can reach the size of a cat.

Attempting to create a familiar fauna around the foreign homes of expatriates, "Acclimatization Societies" active between 1860 and 1890 brought European species to colonial holdings. Members of the American Acclimatization Society in New York were intent on introducing all the species in Shakespeare's works to the New World. His line in *Henry IV,* "Nay, I'll have a starling shall be taught to speak nothing but 'Mortimer'," inspired them to release European starlings (*Sturnus vulgaris*) into Central Park in New York City. Other projects included introduction of the house sparrow (*Passer domesticus*) around the world; importation of various game animals to Australia and New Zealand; introduction of Australian possums to New Zealand; and supply of European carp and trout to a number of river drainages worldwide.

It is evident that humans have been extremely active in moving species from place to place and even to very remote islands. But species also move from place to place as part of a natural process. Islands created by volcanoes in the middle of the oceans have plants and animals. The issue is, How much greater is this human-mediated introduction of colonizers than the "normal" rate? We have seen that the natural colonization rate to distant islands is on the order of thousands of years.

For example, the novel biota of one set of remote islands of the western coast of South America, the Galápagos, became famous as the epiphany for Charles Darwin's *Origin of Species.* A remarkable assemblage of birds found on the Galápagos are called Darwin's finches, *Geospiza,* now generally agreed to have thirteen species. The smallish brown birds differ in the shape and size of their bills and in their feeding habits. Darwin's finches appear to have arisen from a single colonization event involving about thirty individuals.[30] Further, the entire flora of the Galápagos is thought to come from slightly more than four hundred events of successfully colonizing plant species. Given the geologic age of the island at 3 million to 5 million years, a successful plant colonization probably occurred only every ten thousand years or so.[31]

Technological societies have greatly accelerated species introduction to islands around the world. The "miner's canary" for this change has been remote oceanic islands. There are several reasons why islands might be relatively more easily invaded than mainland areas. A relatively low diversity of species exists on islands.[32] Thus there are likely to be no competitors for the invading species, and a potential absence of predators, parasites, and diseases of the invader. The potential food of the invader may not have protective adaptations and therefore be relatively easy for the invader to use. Colonizers may arrive on islands to find vulnerable, little-protected food sources. In addition, many islands are disturbed by human occupation, leaving them open to species that are adapted to the new, human-modified conditions. The small size of islands allows successful invaders to spread over the entire island; their effects on established native species tend to be complete.

Islands are often the stopping points for the international shipping trade, which imports species from a variety of places. In distant oceanic islands, such as Hawaii or New Zealand, more than 40 percent of their diversity comprises exotic plant species. Near islands also have high percentages of exotic, invasive species. On continents, the biodiversity of regions with high levels of exchange with other regions (trade, shipping routes, and the like) tend to have a large exotic component. Even at the level of whole continents, exotic invaders can be a substantial part of the flora.

The tendency to lose indigenous species and gain exotic species seems to be developing as a process to turn over species on continents, as well as on distant islands.[33] What are the limits to this process in a world that is becoming increasingly interconnected? The answer depends to a degree on the factors controlling whether a species is common or rare. The penguin story presented this issue as one generating considerable debate among ecologists. The importance of resolving it stems from the observed changes in connectivity among islands and continents.

The size of continents and their species diversity are related. Larger continents have more species. What if the increased continental exchanges of species made the Earth's landmass effectively a large interconnected continent? If the current relationship between continent size and diversity holds, a sizable reduction in biodiversity should take place on the connected Earth. For mammals, this land-area-to-species relationship implies that a thoroughly connected world might have about half as many mammal species (around 2,000 versus the more than 4,000 today) as are now found on the separated land areas of our planet (Figure 44).

Human action is changing the world and reducing global biodiversity. Peter Vitousek and his colleagues have recently developed a broad overview of the role of invasion.[34] Their analysis finds several important pathways, with and without feedbacks, contributing to global change. The human population is growing in size and resource use, which in turn increases the intensity of industry and agricultural practice. Industrial growth drives increases in emissions of greenhouse

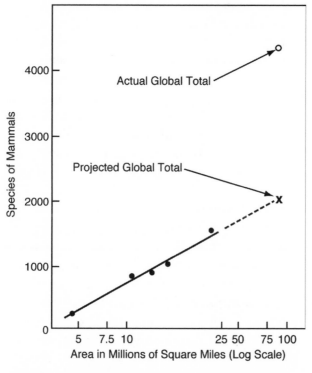

Figure 44. *The relation between size of continent and number of mammal species. The number of species is strongly related to the logarithm of continent size. If the entire terrestrial land surface were united into one supercontinent, the expected diversity of mammals should be about 2,000 species (x on the figure). The actual total number of mammal species is more than twice that number (4,200 species). From P. M. Vitousek et al.,* Introduced species: A significant component of human-caused global change, *New Zealand Journal of Ecology 21 (1997):1–16.*

gases with implications for changing the global climate. Industry and its associated processes alter other cycles of elements as well. Increased agriculture changes the land cover of the Earth with significant implications for global climate and habitat. Other factors, including introduced invasive species and altered climate, reduce biotic diversity. In this web of interactions, change begets change, and loss of diversity is a consequence.

Since this chapter began with the European rabbit, it is appropriate to end with what may be the next phase of human interaction with this invasive species—its eradication. As we consider what should be done with this invasive animal, sympathetic human attitudes toward a docile, peaceable, seemingly harmless creature contrast with another view of the rabbit as a land-destroying displacer of native species and demolisher of crops and pastures. The rabbit is a pet and a pest. It is a game animal and a laboratory animal, a character in children's stories and a source of food and fur.

The history of attempts to control the European rabbit in France demonstrate these contradictory roles. On June 14, 1952, Dr. Armand Delille had become discouraged with the damage done by European rabbits to the crops on his estate, Maillebois, in the département of Eure-et-Loire. He decided to engage in biological warfare. Dr. Delille assumed, incorrectly as it turned out, that the myxoma virus (*Myxomatosis cuniculi*) that he used to knock back the rabbit population would be contained by the walls of his 750-acre (300-ha) estate.[35] He inoculated two rabbits with myxoma virus he had obtained in Switzerland and released them.[36]

The results of his ad hoc experiment were spectacular. By the end of summer in 1952, the disease had spread to nine départements in France. By the end of 1953, rabbits throughout the country were affected, and isolated cases of the disease had appeared in Spain, Belgium, Holland, Germany, and England. Over 90 percent of the French rabbit population was eradicated between 1953 and 1954. Hunters were appalled at the loss of a game animal, and rabbit breeders saw a major threat to their stock and livelihood. In the spotlight of public outrage, in September 1954 Dr. Delille was convicted of illegally spreading an animal disease. He was fined only one franc. Two years later in 1956, he was presented with a medal from the Syndicat National des Forestiers Français in recognition of services rendered to agriculture and silviculture. The increased agricultural and silvicultural productivity resulting from the destruction of the rabbits was valued at a billion francs.[37]

Myxoma is spread between rabbits by a variety of animal vectors—lice, ticks, mites, and some biting insects—but the principal means is fleas and mosquitoes. Although the disease produces benign fibroid lesions in most species of rabbits and hares, it is an entirely different and lethal disease in the European rabbit. In this animal myxoma spreads from the point of the inoculating bite (which initially creates only a small lump) through the animal's tissue in about four days. Conjunctive tissue in the eye begins to swell on the fifth day, then swelling moves to other parts of the body. The animal may die ten days after transmission. Condition, age, and environmental conditions moderate these responses to some degree.[38]

In Australia, as early as 1919, Dr. H. deBeaurepaire Aragão, who had worked with outbreaks of myxomatosis in South America, proposed to the Australian government that a deliberate introduction of the disease could serve to control the European rabbit.[39] The idea was rejected on the grounds of potential revenue loss from the sale of rabbit meat and fear of releasing a virulent disease organism into Australia. Nevertheless, by 1951 myxoma virus had been applied in Australia as a biological control agent against the European rabbit. The result was massive rabbit population reductions on a continental scale.

Public fear of such virulence was fanned by a simultaneous outbreak of viral encephalitis (in humans) in Mildura, a town on the Murray River in Victoria. Mosquitoes spread both, but the two diseases are caused by very different viruses—a subtlety that was quickly lost in the spread of public rumor and general alarm. Concerns were allayed by the announcement in Parliament that three scientists (Sir Ian Clunies Ross, director of the Commonwealth Scientific and Industrial Research Organisation, or CSIRO; Sir Macfarland Burnet, and Professor F. Fenner) had been inoculated with the myxoma virus without harm several months earlier.[40]

Within two years of its introduction, myxoma virus had eliminated up to about 99 percent of Australia's rabbit population. Within two years, the rabbits and the virus began to coevolve. Today, most populations of rabbits have an increased resistance to the disease even though the virus strains remain highly virulent. A net moderation of the disease has resulted.[41]

The Australian government continued to find new ways to make myxoma more effective. In 1969, the European rabbit flea (*Spilopsyllus cunicula*) was brought to Australia to increase vectoring of the disease among rabbits. In 1993, the Spanish rabbit flea (*Xenopsylla cunicularis*), which can survive in drier areas than the European rabbit flea, was introduced for a similar reason. The myxoma virus is a major limiter of the population density of rabbits in Australia, but its current effects are less pronounced than those after initial release of the disease. The population densities of rabbits in high-rainfall areas were likely to have been 5 percent of the predisease levels and 25 percent of the predisease levels in rangeland areas.[42] After the introduction of rabbit hemorrhagic disease (discussed below), the numbers of rabbits were further reduced.

The precipitous decline of rabbits in Australia after the initial release of myxoma virus in 1950–1951 provides some insight into the landscape scale effects of the species. Rates of siltation of dammed ponds and lakes decreased with the renewal of watershed cover. Production of cattle and sheep increased as much as 40 percent in some areas. Despite this recovery, the country's production losses from 1995 on caused by the European rabbit are estimated to be $A600 million per year.[43]

The initial success of myxoma virus as a biological control agent in Australia, and the possibility that the rabbit might develop resistance, inspired interest in other rabbit-specific diseases. A second virus causing rabbit hemorrhagic disease was initially recorded in China in 1984.[44] Rabbit hemorrhagic disease has no effect for the first day after infection, but the rabbits subsequently develop a high temperature and become listless. Within thirty to forty hours of infection, the animals die quietly with no indication of distress. Cause of death appears to be heart failure or lack of oxygen.[45] Rabbits less than eighteen days old survive and develop antibodies to the disease. The susceptibility of young rabbits increases with age. Rabbits twelve weeks of age and older have the same vulnerability as adults.

In 1991 CSIRO obtained permission to import a strain of rabbit hemorrhagic disease virus into Australia. The virus was being tested for its potential as a rabbit control agent on Wardang Island off the coast of South Australia when it escaped to the mainland on October 12, 1995. By May 1996, it had spread to all the states and territories of mainland Australia, and subsequently to Tasmania as well. In October of that year, deliberate releases were begun as a means of controlling European rabbits.

The introduction of rabbit hemorrhagic disease appears to have reduced the populations as much as 90 percent in regions where rabbits were formerly abundant. Vectors different from those of the myxoma virus spread the disease. Rabbit hemorrhagic disease appears to be spread largely by flies. It is prone to epidemics that differ from the myxoma virus in timing and in location because of a complex combination of the biology of the rabbits, the biology of the vectors, and the amount of virus in a population. Initial indications suggest an improvement in the condition of pastures and the associated sustainability of agriculture, increased regeneration of important native plants that are affected by rabbits, a decrease in rabbits and foxes, and an increase in midsized native marsupials.[46] Further, the purchase and application of rabbit poisons have dropped in a number of the areas surveyed.

In the 1980s, several disparate events came together to create an animal control program called virus-vectored immunocontraception (VVI). Wildlife biologists investigated the genetics of the myxoma virus, hoping to modify its severity in the European rabbit. Independently, human medical research developed the concept of using the body's immune system to suppress human fertility. Veterinary researchers found that nonlethal viruses with extra inserted genes could promote antigen responses to the proteins from the extra genes. Therefore, a genetically modified disease inoculated or fed to animals would produce immunity to both the disease and the proteins produced by the inserted genes. If the inserted genes found only sperm cells, for example, the infected animal could recover from a mild disease and have autoimmunity for its reproductive systems. The idea behind VVI was that a rabbit could catch a head cold and wake up the next morning healthy but sterile.

Research in Australia has begun to develop means of VVI for the European rabbit, the European red fox (*Vulpes vulpes*), and the house mouse (*Mus musculus*).[47] The concept is complex but, if successful, the result would be a bioengineered disease that would render the target animals unable to reproduce. The steps required (Figure 45) are all within our current capacity in molecular biology.

The details of how VVI might actually be accomplished in a specific animal present a research endeavor of considerable proportion. Clearly, one would need to be sure that the virus-vectored immunocontraception affects only the target animal. This stipulation colors the choice of proteins to be used in the genetic screening, the selection of the virus to be engineered to induce the immunity, and the manner in which the virus is transferred among animals. An ethical issue might involve the

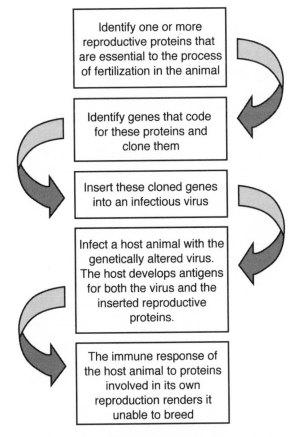

Figure 45. The steps involved in developing virally vectored immunocontraception in animals as a method of biological control. Derived from C. K. Williams, Development and use of virus-vectored immunocontraception, Reproduction, Fertility and Development 9 (1997):169–178.

virulence of the virus. In rabbits, the likely candidate is the myxoma virus: it is host specific, it is delivered to the hosts in a manner that allows it to spread to remote areas (the disease is already present in Australia). The potential proteins to be targeted are those involved with sperm or egg cells.

A perfected VVI system should be inexpensive. It would not involve the use of poisons or traps. It would not pollute. It is humane, in that VVI would not kill the living target animals, but would block the production of new animals. Still, VVI is not without difficulties. Beyond the necessary research to deliver the steps certain ecological considerations are not well understood.

The same is true of social considerations. A Cooperative Research Center has been founded in Australia to tackle the complex issue of biological control of rabbits and other pest animals.[48] Significantly, this unit has included active outreach to the complex constituencies interested in biological control and virus-vectored immunocontraception—including representatives from the sciences, farmers and grazers, and animal-rights activists.

Not unexpectedly, VVI has limitations. It is intended for use as a tool that will be integrated with other biological controls such as rabbit hemorrhagic disease and the strategic use of poisons, along with mechanical destruction of rabbit warrens.

With the current progress in molecular biology, and the tools such progress provides, the potential for genetics to be used to control populations will surely increase. Already mapping the genes of humans, and the subsequent mapping of the genes of both malaria and the *Anopheles* mosquito, are creating the possibility of developing genetically modified mosquitoes to control malaria.[49]

The rabbit is but one example of the invasive potential of human-altered exotics. The level of physical, biological, and chemical warfare that has been raised against rabbits in a battle for control of the productivity of land is an indicator of current capabilities in the elimination of species. The book *Jurassic Park*[50] (and the subsequent movies) raised the possibility of reconstructing extinct species by means of modern molecular biology. An equally feasible and more likely scenario is using biotechnology to eliminate or at least to strongly control the abundance of living species. In a sense, we have done this already for the virulent human disease smallpox, which is extinct as a wild organism. We are hoping to accomplish the same victory for polio. Coincidentally, many of the early researchers on control of the rabbit via myxoma were also active participants in the elimination of smallpox and in developing a polio vaccine.

Unlike deadly diseases such as smallpox or polio, the rabbit is a gentle and seductive invader. But it has proved to be an extremely difficult foe for those seeking to control and eliminate it from occupied landscapes. Even with technology and science aligned against it, the rabbit remains a presence in Australia—as well as in hundreds of other locations that it has successfully invaded.

Human transportation systems have moved species about the world and vastly increased the colonization rate of exotic species. Landscapes altered in structure, biotic diversity, and environment shift away from the species initially located there and toward other species. Common local species can disappear (remember the passenger pigeon), and other local species, even those that were formerly rare, can become abundant to the level of being pests (recall the quelea).

In pronounced cases of landscape modification, change is more and more likely to disadvantage and eliminate native species. At the same time, domesticated animals have been intentionally selected to grow and survive well on human-altered landscapes. As is clear from the European rabbit, these species can be remarkably difficult to remove or control when they become feral or invasive in new locations. In many parts of the world, we are replacing a native biotic diversity with exotic, invasive flora and fauna that are well adapted to humans and to evading our attempts at biological control.

11 *Planetary Stewardship*

Then God said, Let us make man in our image, after our likeness; and let them have dominion over the fish of the sea, and over the fowl of the air, and over the cattle, and over all the earth, and over every creeping thing that creepeth upon the earth.—Genesis 1:26

The biblical admonition to humans to have dominion over the Earth and its creatures has analogies in the tenets of many peoples. The Masai of Kenya believe that they are placed on the world to take care of all the cattle. Ancient Egyptians performed elaborate rituals built around their calendars and intended to ensure the coming of the seasons, the flooding of the Nile, and the moderating of the weather. As our species, *Homo sapiens,* has spread and become more numerous, humankind has altered the face of the planet. Another manifestation of the biblical instruction to have dominion over the Earth is the domestication of animals from the wild condition.

Charles Darwin's cousin, Francis Galton, in 1865 wrote a classic evaluation of the criteria by which animals can be domesticated. He saw domestication as arising from symbiotic interactions between humans and the species to be domesticated. This symbiosis included an appreciation of the companionship afforded by pets. Galton elaborated the conditions for domestication in the following way: "I will briefly restate what appear to be the conditions under which wild animals may become domesticated; 1, they should be hardy; 2, they should have an inborn liking of man; 3, they should be comfort loving; 4, they should be found useful to the savages; 5, they should breed freely; 6, they should be easy to tend." Then, he predicted, "it would appear that every wild animal has had its chance of being domesticated, that those few which fulfilled the above conditions were domesticated long ago, but that the large remainder, who fail sometimes in only one particular, are destined to perpetual wildness so long as their race continues. As civilisation extends they are doomed to be gradually destroyed off the face of the earth as useless consumers of cultivated produce."[1]

Galton's prediction is one potential realization of the biblical instruction to subdue the earth and have dominion over its other living inhabitants. Ecologists and biblical scholars have long debated whether at its root the word "dominion" means "stewardship," with an implication of taking care of the Earth, or "domination," with the sense of controlling and conquering the Earth. The issue of man as being "in nature" (as a part of natural systems) or "out of nature" (as a controller) is frequently raised.

We have dominated the plants and animals of the Earth by domesticating them in some cases, and by extirpating them in others. Galton's prediction of a future world where species that are not domesticated or otherwise useful will be eliminated from a planet filled with cultivated plants seems particularly chilling. Given what has happened since Galton's prediction of the demise of wild creatures in the face of expanding civilization, he probably would be surprised to find his prophecy coming true as rapidly as it seems to be.

I cannot offer simple alternatives. We understand enough about how natural landscapes function to know our alterations of these landscapes are likely to carry us toward Galton's predicted future. Thanks to generations of ecological researchers, we know a lot about the ecology, behavior, and biology of animals, plants, and ecosystems. With active research programs in agriculture and forestry under way around the world, our understanding of the landscapes that we manage for production of food and fiber is probably even better. However, our knowledge pales in the face of what we need to know.

Changes to the Earth from large-scale human alteration are older than our species itself. The human capability to modify the planet's surface probably began long ago with the use of fire by *Homo erectus*.[2] However, our species has increased both planetary change and planetary stewardship. Today our technological society alters the land, the ocean, and the atmosphere with ever greater speed and scale. In the process, we may be modifying the climate as well.

Only a few years ago, climate was thought to be the primary factor influencing the geographic patterns of the soil, the vegetation, and the fauna, but climate was only weakly influenced by these factors in return.[3] An exception was that vegetation altered climate at a very local level—witness the effect of a large tree on the area immediately below its crown. Indeed, the interaction of trees and microclimate helped us to understand vegetation as a dynamically changing mosaic (Chapter 2). These small-scale examples notwithstanding, on a larger scale climate was the determining factor—or so we thought.

The domination of climate over vegetation in large-scale dynamic interactions has been challenged by our improved understanding of the feedback from vegetation to the atmosphere. This "new" appreciation has deep roots. Historical figures such as Christopher Columbus and Noah Webster observed that changes in land cover caused changes in climate. Columbus believed that the presence of forests on the newly discovered islands of the West Indies caused them to have more rainfall than the deforested Azores and Canary Islands.[4] Computer models of the effects of deforestation of tropical islands, analyzed almost five hundred years after Columbus' initial observation, indicate that forested islands should have about three times as much rainfall as the equivalent deforested islands.[5]

Noah Webster felt that the winters in New England became more variable after Europeans deforested North America for purposes of agriculture.[6] In 1853, Antoine-César Becquerel postulated that the presence of forests, and indeed of vegetation in general, affected climate.[7] Becquerel wrote in the middle of the nineteenth century, when the development of agriculture and associated land clearing was radically altering the land surface of Europe and the rest of the world. The impressions of these and other historical figures have been borne out: observational studies demonstrate that the significant amounts of water transpired by plants retard nighttime cooling (both by reducing the outgoing terrestrial radiation and by elevating the dew-point temperature).[8] Some of the chemicals released by plants have been shown to act as particles that promote rainfall. Other chemicals reduce nighttime cooling by radiating heat from the surface of the ground into space.[9]

The question arises, How much do our vegetation changes affect climate? Are we altering more than just the vegetation through our clearing and conversion of landscapes? There have been two different types of efforts by computer-model builders to better understand the effects on climate of changes in the earth's surface. The first type, "sensitivity studies," involves analysis of complex computer models of the atmosphere (called general circulation models, or GCMs) to determine how much the model results change in relation to the way the vegetated surface is represented. The second type focuses on the development of models with processes that better represent the land surface and better predict the consequences of land surface change.

One of the earliest and most interesting sensitivity studies was an investigation of the effects on climate of forest clearing in Amazonia.[10] Part of the motivation for such research is that the Amazon River basin is being cleared of rain forest and converted into farmland and eventually pasturage. A computer-model experiment was developed to determine the effects on the global climate of converting the physical surface conditions over a large area of Amazonia from forest to grassland.[11] In the computer simulation, the removal of forest cover rippled through the global climate to increase the Amazonian temperature by 5.5 degrees Fahrenheit (3 degrees Centigrade) in the deforested region. Given that tropical forest clearing in Amazonia (as elsewhere) is proceeding at a substantial rate, it is easy to see that we need to better understand the influence of the terrestrial surface on the atmosphere.[12]

Subsequent work on the consequences of Amazonian clearing used alternative or improved models of the effects of deforestation over a large area of the Amazon Basin. Comparisons across all of these studies converge in predicting a change to hotter and drier conditions. We have seen the magnitude of the temperature increase. Water from evaporation is increased as much as 20 inches (500 mm), and precipitation is reduced as much as 25 inches (640 mm). These climate changes are large enough to alter the vegetation.

Model experiments in Amazonia indicate potentially significant feedback between the vegetation and the climate. Vegetation alteration in other regions may also be expected to disrupt climate. A primary concern is that such climate changes may alter landscapes irreparably.

When we examined the progressive degrading of the vegetation surrounding the ancient city of Petra (Chapter 9), it was not clear how easily the shrubby, sparse grassland now surrounding the ruins could be recovered to the Mediterranean forest of Petra's past. How much more difficult would this be if the changes in vegetation interacted with atmospheric processes to make the climate less hospitable? When this situation occurs, the feedback between landscape cover and climate is characterized as a destabilizing feedback—the effects of the feedback are to move the ecosystem irreversibly away from its present condition toward some new state. Among others that have been investigated, the Amazon case could involve such feedback between landscape and climate.

As one example, the conversion of sparsely vegetated, semiarid shrubland to desert represents a substantial change in the terrestrial surface and its relation to the atmosphere. A substantial part of the vegetated cover of the sub-Saharan zone called the Sahel is annual grasses (Chapter 6). These grasses sprout each year from the rains brought by the wet season. In the dry season, the Sahelian grasses persist as seeds. Drought delays the germination of the annual grasses and promotes a desert-like condition. Overgrazing can greatly reduce the grass cover as well. In climate-model studies, an overgrazed, bare Sahel interacts with the atmosphere to reduce the rainfall. Devegetating the Sahel delays the rainy season by one month, which shifts the Sahel toward more desert-like conditions.[13] The climate-land interaction converts grassland to desert. More desert changes the climate to promote additional change to desert, and so on. This cycle has been termed desertification.

As a second example, clearing the boreal forest worldwide could potentially increase the reflection of heat from the surface and cause colder air temperatures. In a computer simulation, clearing the great northern evergreen forests produced global cooling.[14] The substantial cooling reduced the temperatures of the entire Northern Hemisphere. Results of this particular study were summarized as follows: "The summer cooling caused by deforestation is sufficient to prevent forest regrowth in much of the deforested area. Thus, boreal deforestation may initiate a long-term irreversible feedback in which the forest does not recover and the tree line moves progressively farther south."

The moa's story in Chapter 8 introduced human-mediated change of ecosystems. Polynesians and other colonizing people (including Europeans) have had profound effects on the biota of islands. Today we have evidence that humans have modified regional landscapes for agriculture all over the world. For the past 200 years and more, scientists have explored the possibility of land modification and clearing as a cause for local or regional climate change. Improved understanding of how the atmosphere interacts with the terrestrial surface has indicated that the interpretations of the past may well be correct. What we do not know is whether these changes are easily reversible. If we criticize past societies for causing extinctions when they lacked the ability to predict the consequences of their actions, what can we say of ourselves if we make wholesale changes to the terrestrial surface *despite* our knowledge that we may well be altering the planet's function?

Let us hope that our planetary stewardship will not take us to the situation envisioned by Francis Galton—a world with landscapes of agriculture and animal husbandry only, with the wild animals removed as useless consumers of cultivated produce, and with a menagerie of rats and rabbits replacing the wild things. We must learn how our planet works before we lose too many of its components.

Our species appears to have wrought change on the essential dynamics of the environment that supports both us and the biotic diversity of our planet. Even from outer space, our planet shows the marks of our stewardship. At our current population density, we are committed to actively managing the planet on which we live; we have no planet to which to retreat should we err.

Given the rate at which we are currently altering the world, we must be strongly motivated to learn much more, and learn it much more rapidly. As we change the landscapes across the face of the Earth, we must understand the consequences. These are challenges of profound importance. Ultimately, they involve the survival of our civilization, our species, and our planet.

Notes

Chapter 1. Introduction

1. I am grateful to Carleton Ray for pointing out the timeliness of Petruchio's questions in this era of species extinctions and introductions. See G. C. Ray, Petruchio's paradox: The oyster or the pearl (pp. 128–136), in J. Cracraft and F. T. Grifo, eds., *The Living Planet in Crisis: Biodiversity and Policy* (New York: Columbia University Press, 1999).

2. C. Wistar, An account of the bones deposited, by the President, in the museum of the society, and represented in the annexed plates, *Transactions of the American Philosophical Society* 4 (1799):525–531.

3. T. Jefferson, A memoir on the discovery of certain bones of a quadruped of the clawed kind in the western part of Virginia, *Transactions of the American Philosophical Society* 4 (1799):246–260. Quotation on p. 255.

4. D. M. Raup and J. J. Sepkoski, Periodicity of extinctions in the geologic past, *Proceedings of the National Academy of Science, U.S.A. Physical Science* 81 (1984): 801–805; D. Jablonski, Background and mass extinctions: The alternation of macroevolutionary regimes, *Science* 231(1986):129–133. Also see review in J. B. Harrington, Climatic change: A review of causes, *Canadian Journal of Forest Research* 17 (1987):1313–39.

5. D. H. Erwin, J. W. Valentine, and J. J. Sepkoski, A comparative study of diversification events: The early Paleozoic versus the Mesozoic, *Evolution* 41 (1987):365–389; M. Jenkins, Species extinctions (pp. 192–205), in B. Groombridge, ed., *Global Biodiversity: Status of the Earth's Living Resources* (London: Chapman and Hall, 1992).

6. W. W. Alvarez, F. Asaro, and H. V. Michel, Extraterrestrial cause for the Cretaceous-Tertiary extinction, *Science* 208 (1980):1095–1108; R. Ganapathy, A major

meteorite impact on the earth 65 million years ago: Evidence from the Cretaceous-Tertiary boundary clay, *Science* 209 (1980):921–923.

7. R. Lewin, Mass extinctions select different victims, *Science* 231 (1986):219–220.

8. J. B. Pollack et al., Environmental effects of an impact-generated dust cloud: Implications for the Cretaceous-Tertiary extinctions, *Science* 219 (1983):287–289.

Chapter 2. The Big Woodpecker That Was Too Picky

1. This figure was reported to J. T. Tanner by a professional bird collector who had shot every ivory-billed woodpecker in one Florida county.

2. I appreciate the use of this whimsy, borrowed from my colleague T. M. Smith.

3. The implications of the cyclic nature of small-scale forest dynamics were clearly elucidated by W. S. Cooper (in The climax forest of Isle Royale, Lake Superior, and its development. I, *Botanical Gazette* 55 (1)(1913):1–44; II, *Botanical Gazette* 55 (2)(1913): 115–140; and III, *Botanical Gazette* 55 (3)(1913):189–235). See also A. S. Watt's classic paper, Pattern and process in the plant community, *Journal of Ecology* 35 (1947):1–22.

4. A. S. Watt, On the ecology of British beech woods with special reference to their regeneration. II, The development and structure of beech communities on the Sussex Downs, *Journal of Ecology* 13 (1925):27–73.

5. Idem, Pattern and process in the plant community, *Journal of Ecology* 35 (1947):1–22.

6. A. F. Cornet et al., Water flows and the dynamics of desert vegetation stripes (pp. 327–345), in A. J. Hansen and F. DiCastri, eds., *Landscape Boundaries: Consequences for Biotic Diversity and Ecological Flows* (Berlin: Springer-Verlag, 1992).

7. Africa: S. B. Broaler and C. A. H. Hodge, Observations on vegetation arcs in the northern region, Somali Republic, *Journal of Ecology* 52 (1964):511–544; C. F. Hemming, Vegetation arcs in Somaliland, *Journal of Ecology* 53 (1965):57–68; L. P. White, *"Brousse tigrée"* patterns in southern Niger, *Journal of Ecology* 58 (1970):549–553; L. P. White, Vegetation stripes on sheet wash surfaces, *Journal of Ecology* 59 (1971):615–622. North America: C. Montaña, J. Lopez-Portillo, and A. Mauchamp, The response of two woody species to the conditions created by a shifting ecotone in an arid ecosystem, *Journal of Ecology* 78 (1990):789–798; C. Montaña, The colonization of bare areas in two-phase mosaics of an arid ecosystem, *Journal of Ecology* 80 (1992):315–327; A. F. Cornet et al., Water flows and the dynamics of desert vegetation stripes (pp. 327–345), in Hansen and DiCastri, *Landscape Boundaries.* Australia: R. O. Slatyer, Methodology of a water balance study conducted on a desert woodland community in central Australia (pp. 15–26), in UNESCO *Arid Zone Research 16: Plant-Water Relationships in Arid and Semi-arid Conditions* (Paris: UNESCO, 1961); G. Pickup, The erosion cell—a geomorphic approach to landscape classification in range assessment, *Australian Rangeland Journal* 7 (1985):114–121; D. J. Tongway and J. A Ludwig, Vegetation

and soil patterning in semi-arid mulga lands of eastern Australia, *Australian Journal of Ecology* 15 (1990):23–34.

8. C. Montaña, Las foraciones vegetales (pp. 167–198), in C. Montaña, ed., *Estudio Integrado de los Recursos Vegetación, Suelo Y Aqua en la Reserva de la Biosfera de Mapimí,* (Xalapa: Instituto de Ecologia, 1988).

9. Idem, The colonization of bare areas in two-phase mosaics of an arid ecosystem, *Journal of Ecology* 80 (1992):315–327.

10. D. G. Sprugel, Dynamic structure of wave-generated *Abies balsamea* forests in northeastern United States, *Journal of Ecology* 64 (1976):889–911.

11. Y. Oshima et al., Ecological and physiological studies on the vegetation of Mt. Shimagaree. I, Preliminary survey of the vegetation of Mt. Shimagaree, *Botanical Magazine of Tokyo* 71 (1958):289–300. Quotation on p. 289.

12. Sprugel, Dynamic structure.

13. T. C. Whitmore, On pattern and process in forests (pp. 45–59), in E. I. Newman, ed., *The Plant Community as a Working Mechanism* (Oxford: Blackwell Scientific Publications, 1982); N. V. L. Brokaw, Gap-phase regeneration in a tropical forest, *Ecology* 66 (1985):682–687; idem, Treefalls, regrowth, and community structure in tropical forests (pp. 101–108), in S. T. A. Pickett and P. S. White, eds., *The Ecology of Natural Disturbance and Patch Dynamics* (New York: Academic Press, 1985); J. Silvertown and B. Smith, Gaps in the canopy: The missing dimension in vegetation dynamics, *Vegetatio* 77 (1988):57–60; R. A. A. Oldeman, *Forests: Elements of Silvology* (Berlin: Springer-Verlag, 1991).

14. H. H. Shugart, *A Theory of Forest Dynamics: The Ecological Implications of Forest Succession Models* (New York: Springer-Verlag, 1984); idem, Dynamic ecosystem consequences of tree birth and death patterns, *BioScience* 37 (1987):596–602.

15. G. S. Hartshorn, Tree falls and tropical forest dynamics (pp. 617–638), in P. B. Tomlinson and M. H. Zimmermann, eds., *Tropical Trees as Living Systems* (Cambridge: Cambridge University Press, 1978).

16. F. Hallé, R. A. A. Oldeman, and P. B. Tomlinson, *Tropical Trees and Forests* (Hiedelberg: Springer Verlag, 1978).

17. Oldeman, in *Forests,* points out that forest dwellers in various parts of the world have a rich vocabulary to describe processes and patterns of tree falls, patches in forests, and the like. He notes *chablis* and *volis* in French, *rytä* in Finnish, and *traa* and *loo* in Dutch, as such terms. Oldeman speculates that the variety of grape used to produce chablis wine may have originated from a variety of *Vitis vinifera* found in small forest openings, or chablis in older French idiom.

18. R. B. Foster, *Tachigalia versicolor* is a suicidal neotropical tree, *Nature* 268 (1977): 624–626.

19. T. Hardy, *Wessex Poems and Other Verses* (New York: Harper, 1898).

20. H. H. Shugart, *Terrestrial Ecosystems in Changing Environments* (Cambridge: Cambridge University Press, 1998).

21. F. H. Bormann and G. E. Likens, *Pattern and Process in a Forested Ecosystem,* (New York: Springer-Verlag, 1979). Quotation on p. 175.

22. T. C. Whitmore, On pattern and process in forests (pp. 45–59), in E. I. Newman, ed., *The Plant Community as a Working Mechanism* (Oxford: Blackwell Scientific Publications, 1982). Quotation on p. 45.

23. Tropical rain forests: A. Aubréville, La forêt colonaile: Les forêts de l'Afrique occidentale française, *Annales de l'Academie des Sciences Coloniale* 9 (1938):1–245. Translated by S. R. Eyre, Regeneration patterns in the closed forest of Ivory Coast (pp. 41–55), in S. R. Eyre, ed., *World Vegetation Types* (London: Macmillan, 1991); E. W. Jones, Ecological studies on the rain forest of southern Nigeria. IV, The plateau forest of the Okumu forest reserve, *Journal of Ecology* 43 (1955):564–594, and 44 (1956):83–117; T. C. Whitmore, Change in time and the role of cyclones in the tropical rain forest on Kolombangara, Solomon Islands, *Commonwealth Forestry Institute Paper* 46 (1974); D. H. Knight, A phytosociological analysis of species-rich tropical forest on Barro Colorado Island, Panama, *Ecological Monographs* 45 (1975):259–284; G. S. Hartshorn, Tree falls and tropical forest dynamics (pp. 617–638), in P. B. Tomlinson and M. H. Zimmermann, eds., *Tropical Trees as Living Systems* (Cambridge: Cambridge University Press, 1978). Temperate forests: E. W. Jones, The structure and reproduction of the virgin forests of the north temperate zone, *New Phytologist* 44 (1945):130–148; H. M. Raup, Some problems in ecological theory and their relation to conservation, *Journal of Ecology* 52 (Suppl.) (1964):19–28; P. S. White, Pattern, process and natural disturbance in vegetation, *Botanical Review* 45 (1979):229–299; C. D. Oliver, Forest development in North America, *Forest Ecology and Management* 3 (1981):153–168; G. F. Peterken and E. W. Jones, Forty years of change in Lady Park Wood: The old-growth stands, *Journal of Ecology* 75 (1987):477–512. For review and discussion, see also Whitmore, On pattern and process.

24. O. Rackham, Mixtures, mosaics and clones: The distribution of trees within European woods and forests (pp. 1–20), in M. G. R. Cannell, D. C. Malcolm, and P. A. Robertson, eds., *The Ecology of Mixed-Species Stands of Trees* (Oxford: Blackwell Scientific Publications, 1992).

25. Shugart, *Terrestrial Ecosystems.*

26. It takes about forty years in most cases.

27. Temperate forest examples: Jones, Virgin forests; Rackham, Mixtures, mosaics and clones. Tropical forest examples: M. D. Swain and J. B. Hall, The mosaic theory of forest regeneration and the determination of forest composition in Ghana, *Journal of Tropical Ecology* 4 (1988):253–269.

28. J. T. Tanner related from his experience at the Singer tract in Louisiana that, in the early spring, the new leaves filling the south side of large tree canopies faster than elsewhere caused enough of a weight difference to pull over large trees that were rooted in the soft wet soil. He recalled occasionally hearing a crash from the fall of

a giant canopy tree on still days, and the disquieting effect this event had on the singing of the birds (personal communication).

Chapter 3. The Black-Headed Bird Named Whitehead

1. The living species are emperor penguin, *Aptenodytes forsteri;* king penguin, *A. patagonicus;* adelie penguin, *Pygoscelis adeliae;* gentoo penguin, *P. papua;* chinstrap penguin, *P. antarctica;* rockhopper penguin, *Eudyptes chrysocome;* macaroni penguin, *E. chrysolophus;* royal penguin, *E. schlegeli;* fiordland crested penguin, *E. pachyrhynchus;* erect-crested penguin, *E. sclateri;* Snare's Island penguin, *E. robustus;* yellow-eyed penguin, *Megadyptes antipodes;* fairy penguin, *Eudyptula minor;* magellanic penguin, *Spheniscus magellanicus;* humboldt penguin, *S. humboldti;* African penguin, *S. demersus;* Galápagos penguin, *S. mendiculus.* J. del Hoyo, A. Elliott, and J. Sargatal, eds., *Handbook of the Birds of the World,* vol. 1, *Ostrich to Ducks* (Barcelona: Lynx Edicions, 1992). There are eighteen species, if one recognizes the white-flippered variety of fairy penguin, *Eudyptula albosignata,* as a separate species.

2. Del Hoyo, Elliott, and Sargatal, *Handbook of the Birds.*

3. *Alca pica* is actually no longer in use as the scientific name of the razorbill (*Alca torda*).

4. Niche was formalized as a scientific concept in 1917: J. Grinnell, The niche relations of the California thrasher, *Auk* 34 (1917):364–382. The etymology of "niche" was discussed in P. M. Gaffney, The roots of the niche concept, *American Naturalist* 109 (1975):490, and D. L. Cox, A note on the queer history of "niche," *Bulletin of the Ecological Society of America* 61 (1980):201–202. These authors point out that in Grinnell's time it was not a novel word, even in an ecological context. Thoreau, for example, used it without feeling the necessity for elaboration. Also see J. Grinnell, The origin and the distribution of the chestnut-backed chickadee, *Auk* 21 (1904):364–382, and idem, Geography and evolution, *Ecology* 5 (1924): 225–229.

5. Grinnell, Niche relations; and idem, Field tests and theories concerning distributional control, *American Naturalist* 51 (1917):115–128.

6. Grinnell, in Origin and distribution, also discussed many of the concepts of how animal distributions are controlled.

7. J. Grinnell, Presence and absence of animals, *University of California Chronicles* 30 (1928):429–450. Quotation on p. 429.

8. L. G. Ramensky, Vestnik opytnogo dela Sredne-Chernoz [Basic lawfulness in the structure of vegetation cover], excerpted in E. J. Kormandy, ed., *Readings in Ecology,* pp. 151–152 (Englewood Cliffs, N.J., Prentice-Hall, 1924); H. A. Gleason, The individualistic concept of plant association, *Bulletin of the Torrey Botanical Club* 53 (1926):1–20.

9. F. E. Clements, *Plant Succession: An Analysis of the Development of Vegetation* (Washington, D.C.: Carnegie Institute, 1916); F. E. Clements, *Plant Succession and Indicators* (New York: Wilson, 1928).

10. The presence of bitter or poisonous pasture weeds indicates overgrazing. One can also determine the presence of mineral deposits from the location of species tolerant of various heavy metals.

11. For example, Whittaker and his colleagues felt that Grinnell's concept emphasized the habitat sufficiently that niche was synonymous with habitat. They referred Grinnell's niche as "the habitat or place niche." R. H. Whittaker, S. A. Levin, and R. B. Root, Niche, habitat and ecotope, *American Naturalist* 109 (1973):479–482.

12. Autecological studies focus on the relations between individuals and their environment; synecological studies (which we shall discuss later) emphasize the relations between assemblages of multiple species or ecosystems and the environment.

13. J. Grinnell, The designation of birds' ranges, *Auk* 44 (1927):322–325.

14. See F. C. James et al., The Grinnellian niche of the wood thrush, *American Naturalist* 124 (1984):17–47.

15. R. O. Erickson, The *Clematis fremontii* var. *riehlii* population in the Ozarks, *Annals of the Missouri Botanical Garden* 23 (1945):413–461.

16. Dolomite is a common mineral composed of calcium magnesium carbonate, $CaMg(CO_3)_2$.

17. The data for this example are as tabulated by Erickson in 1945.

18. F. I. Woodward, *Climate and Plant Distribution* (Cambridge: Cambridge University Press, 1987).

19. Many of these applications involve multivariate statistical procedures and large calibration data sets. The methodologies are complex and ecologists debate the most appropriate procedures to be used. Nonetheless, the methods all attempt to ascertain which measurable features of the environment most reliably predict where and when a given species will occur.

20. J. Rice, R. D. Ohmart, and B. W. Anderson, Bird community use of riparian habitats: The importance of temporal scale in interpreting discriminant function analysis (pp. 186–196), in D. E. Capen, ed., *The Use of Multivariate Statistics in Studies of Wildlife Habitat* (Fort Collins, Colo.: United States Department of Agriculture—Forest Service, 1981); idem, Turnovers in species composition of avian communities in contiguous riparian habitats, *Ecology* 64 (1983):1444–55; idem, Habitat selection attributes of an avian community: A discriminant function analysis, *Ecological Monographs* 53 (1983):263–290; idem, Limits in a data rich model: Modeling experience with habitat management on the Colorado River (pp. 79–86), in J. Verner, M. L. Morrison, and C. J. Ralph, eds., *Wildlife 2000: Modeling Habitat Relationships of Terrestrial Vertebrates* (Madison: University of Wisconsin Press, 1986).

21. Rice, Ohmart, and Anderson, Turnovers in species composition; idem, Habitat se-

lection attributes; B. W. Anderson, R. D. Ohmart, and J. Rice, Avian and vegetation community structure and their seasonal relationships in the lower Colorado River valley, *Condor* 85 (1983):392–405.

22. A similar species, the long-billed thrasher (*Toxostoma longirostre*) is found in southern Texas in the vicinity of Brownsville and has a small zone of overlap with the brown thrasher.

23. C. Elton, *Animal Ecology* (New York: Macmillan, 1927). Quotation on p. 63.

24. P. S. Giller, *Community Structure and the Niche* (London: Chapman and Hall, 1984).

25. J. R. Griesemer, Niche: Historical perspectives (pp. 231–240), in E. F. Keller and E. A. Lloyd, eds., *Keywords in Evolutionary Biology* (Cambridge, Mass.: Harvard University Press, 1992).

26. A. J. Lotka, *Elements of Physical Biology* (Baltimore: Williams and Wilkins, 1925); V. Volterra, Variations and fluctuations of the number of individuals in animal species living together [translation], in R. N. Chapman, ed., *Animal Ecology* (New York: McGraw-Hill, 1926).

27. G. F. Gause, *The Struggle for Existence* (Baltimore: Williams and Wilkins, 1934). Quotation on p. 19. See also idem, The ecology of populations, *Quarterly Review of Biology* 7 (1932):27–46.

28. G. Hardin, The competitive exclusion principle, *Science* 131 (1960):1292–97.

29. J. Grinnell, Geography and evolution, *Ecology* 5 (1924):225–229; C. Elton, *Animal Ecology* (New York: Macmillan, 1927).

30. Both M. Williamson, in *The Analysis of Biological Populations* (London: Edward Arnold, 1972), and W. Arthur, in *The Niche in Competition and Evolution* (New York: John Wiley and Sons, 1987), provide reviews of laboratory and field studies. Evidence for competition as a force working to structure the community was independently reviewed by T. W. Schoener, Field experiments on interspecific competition, *American Naturalist* 122 (1983):240–285, and J. H. Connell, On the prevalence and relative importance of interspecific competition: Evidence from field experiments, *American Naturalist* 122 (1983):661–696. They tabulated papers that claimed to investigate the outcome of experiments on competition among two or more similar species. Connell reviewed papers in six journals (*American Naturalist, Ecological Monographs, Ecology, Journal of Animal Ecology, Journal of Ecology,* and *Oecologia*) published between 1974 and 1982. He identified evidence for competition in about 40 percent of the 527 experiments. Schoener conducted a broader review and had somewhat different criteria for his enumeration. See T. W. Schoener, Some comments on Connell's and my reviews of field experiments on interspecific competition, *American Naturalist* 125 (1985):730–740. About 90 percent of the experiments Schoener tabulated demonstrated competition.

31. M. L. Cody, *Competition and the Structure of Bird Communities* (Princeton: Princeton University Press, 1974).

32. These are the alpha and beta parameters in the Lotka-Volterra equations. See R. Levins, *Evolution in Changing Environments* (Princeton: Princeton University Press, 1968).

33. Gause, *Struggle for Existence;* G. E. Hutchinson, Concluding remarks, *Cold Spring Harbor Symposium on Quantitative Biology* 22 (1957):415–427; J. M. Diamond, Assembly of species communities (pp. 342–344), in M. L. Cody and J. M. Diamond, eds., *Ecology and Evolution of Communities* (Cambridge, Mass.: Harvard University Press, 1975); and Cody, *Competition and Structure of Bird Communities.*

34. R. Levins, Extinction (pp. 77–107), in M. Gerstenhaber, ed., *Some Mathematical Problems in Biology* (Providence: American Mathematical Society, 1970).

35. R. H. MacArthur and R. Levins, The limiting similarity, convergence and divergence of coexisting species, *American Naturalist* 101 (1967):377–385.

36. P. S. Giller, *Community Structure and the Niche* (London: Chapman and Hall, 1984); J. Roughgarden, Resource partitioning among competing species—a coevolutionary approach, *Theoretical Population Biology* 9 (1976):388–424; W. L. Brown and E. O. Wilson, Character displacement, *Systematic Zoology* 5 (1956): 48–64.

37. F. C. James et al., The Grinnellian niche of the wood thrush, *American Naturalist* 124 (1984):17–47.

38. R. H. Peters, *A Critique for Ecology* (Cambridge: Cambridge University Press, 1991).

39. For example, M. Levandowsky, The white queen speculation, *Quarterly Review of Biology* 52 (1977):383–386; E. C. Pielou, *Mathematical Ecology,* 2nd ed. (New York: John Wiley and Sons, 1977).

40. R. H. Peters, Tautology in evolution and ecology, *American Naturalist* 110 (1977): 1–12; idem, *A Critique for Ecology.*

41. For example, H. V. Cornell and J. H. Lawton, Species interactions, local and regional processes, and limits to the richness of ecological communities—A theoretical perspective, *Journal of Animal Ecology* 61 (1992):1–12, note that "there seem to be numerous cases where there is open niche space." They conclude that "many ecological communities should not be saturated."

42. A. Hort, *Enquiry into Plants and Minor Works on Odours and Weather Signs* (Theophrastus, trans. Sir Albert Hort), vols. 1 and 2 (London: W. Heinemann, 1916); A. von Humboldt, *Ideen zu einer Geographie der Pflanzen* (Tübingen: F. G. Cotta, 1807; reprinted 1963 by Wissenschaftliches Buches, Darmstadt); L. R. Holdridge, *Life Zone Ecology* (San Jose, Costa Rica: Tropical Science Center, 1967); E. O. Box, *Macroclimate and Plant Forms: An Introduction to Predictive Modeling in Phytogeography* (The Hague: Dr. W. Junk, 1981).

43. J. B. Wilson, in Mechanisms of species coexistence: Twelve explanations for Hutchinson's "Paradox of the Plankton": Evidence from New Zealand plant communities, *New Zealand Journal of Ecology* 13 (1990):17–42, considers much the

same issue: How do the species of a community persist without the occurrence of competitive exclusion? He finds that, for the flora of New Zealand, it is likely that environmental variation disequilibrates ecosystems before competition can manifest strong effects as a structuring force. Wilson (p. 32) states, "it seems that all plant communities are always in a state of change in response to climate." M. Zobel, in Plant species coexistence—The role of historical, evolutionary and ecological factors, *Oikos* 65 (1992):314–320) also considers factors that structure plant communities. Zobel (p. 318) notes, "The diversity pattern of real plant communities does not suggest that the competitive exclusion of similar species can be the basis for explaining coexistence."

44. C. Barbraud and H. Welmerskitch, Emperor penguins and climate change, *Nature* 411 (2001):183–186.

Chapter 4. The Rat That Hid Time in Its Nest

Epigraph. A. L. Kroeber, *Karok Myths* (Berkeley: University of California Press, 1980).

1. E. R. Hall, *Mammals of North America,* vol. 2 (New York: John Wiley, 1981).
2. T. A. Vaughan, Ecology of living packrats (pp. 14–27), in J. L. Betancourt, T. R. Van Devender, and P. S. Martin, eds., *Packrat Middens: The Last 40,000 Years of Change* (Tucson: University of Arizona Press, 1990).
3. W. G. Spaulding et al., Packrat middens: Their composition and methods of analysis (pp. 59–84), in Betancourt, Van Devender, and Martin, *Packrat Middens.*
4. R. B. Finley, Woodrat ecology and behavior and the interpretation of paleomiddens (pp. 28–42), in Betancourt, Van Devender, and Martin, *Packrat Middens.* K. P. Dial and N. J. Czaplewski, Do woodrat middens accurately represent the animals' environments and diets? The Woodhouse Mesa study (pp. 43–58), in Betancourt, Van Devender, and Martin, *Packrat Middens.* Betancourt and his colleagues provide a number of case studies. In Chapter 9 we discuss the use of hyrax middens in Jordan to date the changes in vegetation near the ancient city of Petra.
5. Theophrastus observed the positive relationship between altitude and latitude with respect to climate and vegetation. See A. Hort, *Enquiry into Plants and Minor Works on Odours and Weather Signs* [Theophrastus' work translated by Sir Albert Hort], vols. 1 and 2 (London: W. Heinemann, 1916); and A. G. Morton, *History of Botanical Science* (London: Academic Press, 1981), for this and other observations of Theophrastus' on vegetation and climate.
6. K. Cole, Late Quaternary zonation of the vegetation in the eastern Grand Canyon, *Science* 217 (1982):1142–45; idem, Past rates of change, species richness, and a model of vegetational inertia in the Grand Canyon, Arizona, *American Naturalist* 125 (1985):289–303.
7. H. J. B. Birks and H. H. Birks, *Quaternary Palaeoecology* (London: Edward

Arnold, 1980); A. C. Ashworth, The response of beetles to Quaternary climate changes (pp. 119–128), in B. Huntley et al., *Past and Future Rapid Environmental Changes: The Spatial and Evolutionary Responses of Terrestrial Biota* (Berlin: Springer-Verlag, 1997).

8. Fossils in ocean: J. D. Hays, J. Imbrie, and N. J. Shackleton, Variations in the Earth's orbit: Pacemaker of the ice ages, *Science* 194 (1976):1121–32; J. Imbrie et al., The orbital theory of Pleistocene climate: Support from revised chronology of marine ^{18}O record (pp. 269–305), in A. Berger et al., eds., *Milankovitch and Climate,* pt. 1 (Dordrecht: D. Reidel, 1984). Ice in glaciers: W. Dansgaard et al., A new Greenland deep ice core, *Science* 218 (1982):579–584; B. Saltzman, Climatic systems analysis, *Advanced Geophysics* 25 (1982):173–233.

9. The cycles of change were proposed by J. Croll, On the eccentricity of the Earth's orbit, and its physical relations to the glacial epoch, *Philosophy Magazine* 33 (1867):119–131, and derived mathematically by M. M. Milankovitch, *Canon of Insolation and the Ice-Age Problem* (Beograd: Köningliche Serbische Academie, 1941) [English translation by the Israel Program for Scientific Translation, published in 1969 by the U.S. National Science Foundation, Washington, D.C.].

10. J. Imbrie and J. Z. Imbrie, Modelling the climatic response to orbital variations, *Science* 207 (1980):943–953; A. Berger, Accuracy and frequency stability of the Earth's orbital elements during the Quaternary (pp. 3–126), in Berger et al., *Milankovitch and Climate,* pt. 1.

11. T. J. Crowley et al., Role of seasonality in the evolution of climate during the past 100 million years, *Science* 231 (1986):579–584.

12. A. Robock, Internally and externally caused climatic change, *Atmospheric Science* 35 (1978):1111–22.

13. Interpretation of the pollen in a lake deposit is complex. Currents stir sediments in some lakes, yet in other lakes one can clearly discern each year's deposition. The various plants emit different amounts of pollen, so that given amounts indicate different levels of presence of different species. Pollen grains differ in size and in the distances that they travel. Thus, finding grains of pollen from one plant species indicates that it was found near the lake, whereas finding the grains of another species, whose pollen travels farther, indicates only it was in the region of the lake. Procedures to interpret pollen data correctly are constantly being improved and tested.

14. D. K. Grayson, Nineteenth-century explanations of Pleistocene extinctions: A review and analysis (pp. 5–39), in P. S. Martin and R. G. Klein, eds., *Quaternary Extinctions* (Tucson: University of Arizona Press, 1984).

15. H. W. Longfellow, *Evangeline,* pt. 1.

16. R. G. West, *Pleistocene Geology and Biology* (London: Longman, 1977).

17. B. Huntley and H. J. B. Birks, *An Atlas of Past and Present Pollen Maps for Europe: 0–13,000 Years Ago* (Cambridge: Cambridge University Press, 1983).

18. R. G. West, Interglacial and interstadial vegetation in England, *Proceedings of the Linnaean Society of London* 172 (1961):81–89; idem, *Pleistocene Geology and Biology;* M. B. Davis, Pleistocene biography of temperate deciduous forests, *Geoscience and Man* 13 (1976):13–26; H. E. Wright, Jr., Quaternary vegetation history—Some comparisons between Europe and America, *Annual Review of Earth Planetary Sciences* 5 (1977):123–158; W. A. Watts, Europe (pp. 155–192), in B. Huntley and T. Webb III, eds., *Vegetation History* (Dordrecht: Kluwer, 1988); H. J. B. Birks, Holocene isochrone maps and patterns of tree-spreading in the British Isles, *Journal of Biogeography* 16 (1989):503–540; H. R. Delcourt and P. A. Delcourt, *Quaternary Ecology: A Paleoecological Perspective* (London: Chapman and Hall, 1991).

19. M. B. Davis, Pleistocene biography of temperate deciduous forests, *Geoscience and Man* 13 (1976):13–26; idem, Quaternary history and the stability of forest communities (pp. 132–153), in D. C. West, H. H. Shugart, and D. B. Botkin, eds., *Forest Succession: Concepts and Application* (New York: Springer-Verlag, 1981); idem, Outbreaks of forest pathogens in Quaternary history, *Proceedings of the Fourth International Palynol Conference Lucknow* 3 (1981):216–227; idem, Quaternary history of deciduous trees of eastern North America and Europe, *Annals of the Missouri Botanical Garden* 70 (1983):550–563; idem, Climatic instability, time lags, and community disequilibrium (pp. 269–284), in J. Diamond and T. J. Case, eds., *Community Ecology* (New York: Harper and Row, 1986).

20. B. Huntley, Europe (pp. 341–383), in Huntley and Webb, *Vegetation History;* Huntley and Birks, *Atlas of Pollen Maps.*

21. H. H. Shugart, R. Leemans, and G. B. Bonan, eds., *A Systems Analysis of the Global Boreal Forest* (Cambridge: Cambridge University Press, 1992).

22. Shugart, Leemans, and Bonan, *A Systems Analysis,* provides a worldwide description of the species of the boreal forests.

23. T. Webb III, Glacial and Holocene vegetation history: Eastern North America (pp. 385–414), in Huntley and Webb, *Vegetation History.*

24. Tropical: T. C. Whitmore and G. T. Prance, eds., *Biogeography and Quaternary History in Tropical America* (Oxford: Oxford Science Publications, 1987); J. Haffer, Quaternary history of tropical America (pp. 1–18), in Whitmore and Prance, *Biogeography and Quaternary History;* T. van der Hammen, South America (pp. 307–340), in Huntley and Webb, *Vegetation History.*

Temperate: Delcourt and Delcourt, *Quaternary Ecology.*

Arctic: H. F. Lamb and M. E. Edwards, The Arctic (pp. 519–555), in Huntley and Webb, *Vegetation History.*

25. P. S. Martin and H. E. Wright, Jr., eds., *Pleistocene Extinctions: The Search for a Cause* (New Haven: Yale University Press, 1967); Martin and Klein, *Quaternary Extinctions.*

26. D. S. Webb, Ten million years of mammal extinctions in North America (pp. 189–210), in Martin and Klein, *Quaternary Extinctions.*

27. Martin and Klein, *Quaternary Extinctions;* R. N. Owen-Smith, *Megaherbivores: The Influence of Very Large Body Size on Ecology* (Cambridge: Cambridge University Press, 1988).

28. See Martin and Klein, *Quaternary Extinctions,* for the various views and three summaries of the available data.

29. T. Nilsson, *The Pleistocene* (Stuttgart: Ferdinand Enke Verlag, 1983); E. Anderson, 1984, Who's who in the Pleistocene: A mammalian beastiary (pp. 5–39), in Martin and Klein, *Quaternary Extinctions.*

30. One subspecies of the elephant found in moist forests is recognized by some as a species separate from the African elephant. This subspecies is smaller, narrower across the shoulders, and has a shorter gestation time.

31. V. J. Maglio, Origin and evolution of the Elephantidae, *Transactions of the American Philosophical Society (New Series)* 63 (1973):1–149; L. D. Agenbroad, New World mammoth distribution (pp. 90–108), in Martin and Klein, *Quaternary Extinctions.*

32. Ibid.

33. Owen-Smith, *Megaherbivores.*

34. P. S. Martin and D. W. Steadman, Prehistoric extinctions on islands and continents (pp. 17–55), in R. D. E. McPhee, ed., *Extinctions in Near Time* (New York: Kluwer Academic/Plenum Publishers, 1999).

35. Ibid.

36. R. D. E. MacPhee and P. A. Marx, The 40,000 year plague: Humans, hyperdisease and first-contact extinctions (pp. 169–217), in S. Goodman and B. Patterson, eds., *Natural Change and Human Impact in Madagascar* (Washington, D.C.: Smithsonian Institution Press, 1997).

37. C. F. Baes, Jr., et al., Carbon dioxide and climate: The uncontrolled experiment, *American Scientist* 65 (1977):310–320; C. D. Keeling, The global carbon cycle: What we know and could know from atmospheric, biospheric and oceanic observations (pages II.3–II.62), in *Proceedings of the CO_2 Research Conference: Carbon Dioxide, Science, and Consensus* (Springfield, Va.: NTIS, 1983).

38. F. I. Woodward, Stomatal numbers are sensitive to increases in CO_2 from pre-industrial levels, *Nature* 327 (1987):617–618.

39. Idem, Plant responses to past concentrations of CO_2, *Vegetatio* 104/105 (1993): 145–155.

40. M. E. Mann, R. S. Bradley, and M. K. Hughes, Northern Hemisphere temperatures during the past millennium: Inferences, uncertainties, and limitations, *Geophysical Research Letters* 26 (1999):759–762.

41. J. L. Sarmiento and M. Bender, Carbon biogeochemistry and climate change, *Photosynthesis Research* 39 (1994):209–234.

42. R. E. Dickinson, How will climate change? (pp. 206–270), in B. Bolin et al., eds.,

The Greenhouse Effect, Climatic Change, and Ecosystems (Chichester: John Wiley, 1986).

Chapter 5. The Earthquake Bird and the Possum

1. C. W. Stover and J. L. Coffman, The New Madrid earthquake of February 7, 1812, in *Seismicity of the United States, 1568–1989 (Revised),* U.S. Geological Survey Professional Paper 1527 (Washington, D.C.: Government Printing Office, 1993).

2. The New Madrid earthquake of December 16, 1811, is the fifth largest in the contiguous United States; that of January 23, 1812, is the eighth largest.

3. A. C. Bent, *Life Histories of North American Wood Warblers,* Smithsonian Institution (Washington, D.C.: Government Printing Office, 1953).

4. The species is described in J. J. Audubon, *Ornithological Biography,* vol. 1, p. 323 (Edinburgh: Adam and Charles Black, 1831). John Bachman and his family were indeed very close to John James Audubon, and Bachman's daughters married two of Audubon's sons.

5. C. S. Galbraith, Bachman's warbler (*Helminthophila bachmani*) in Louisiana, *Auk* 5 (1888):323.

6. O. Widmann, The summer home of the Bachman's warbler no longer unknown: A common breeder in the St. Francis River region of southeastern Missouri and northeastern Arkansas, *Auk* 4 (1887):305–310.

7. The descendants of Dr. John Bachman still reside in Charleston. They pronounce their family name "Backman" (P. B. Hamel and S. A Gauthreaux, personal communication).

8. A. T. Wayne, Bachman's warbler (*Helinthophila bachmanii*) rediscovered in South Carolina, *Auk* 8 (1901):83–84.

9. R. V. Remsen, Was Bachman's warbler a bamboo specialist? *Auk* 103 (1986):216–219.

10. H. M. Stevenson, The recent history of the Bachman's warbler, *Wilson Bulletin* 84 (1972):344–347.

11. W. Brewster, Notes on the Bachman's warbler (*Helminthophila bachmanii*), *Auk* 8 (1891):149–157; Wayne, Bachman's warbler.

12. H. M. Stevenson, who had earlier collected over thirty of the birds, argued that this was at the time a relatively small dent in their continental population. He also saw the Bachman's warbler as a species whose time had come (Stevenson, Recent history). F. W. Chapman, who along with W. Brewster collected about fifty Bachman's warblers in the 1890s (Brewster, Notes), later founded the birding tradition known as the Christmas Bird Count. The event now involves 50,000 participants who count the birds found at 1,800 locations in the Western Hemisphere. The concept was initially proposed by Chapman as an antidote to the traditional "side hunt,"

which by 1900 had only recently fallen into disfavor. The side hunt was a contest in which teams of "sportsmen" competed to see who could shoot the most birds and mammals on Christmas Day. Hundreds of nongame birds were killed in these events, and the scores were published in sporting magazines. While Chapman was strongly opposed to this sort of carnage, he remained a strong supporter of the collection of bird specimens by competent ornithologists as a necessity for taxonomic and biogeographic understanding of birds. (F. M. Chapman, A Christmas bird-census, in F. M. Chapman, ed., *Bird Lore 14* (Harrisburg, Penn.: MacMillan Company, Dec. 1900).

13. The late J. T. Tanner saw several Bachman's warblers in the Singer Tract in Louisiana while surveying for the ivory-billed woodpecker. He commented that his observations of the species were associated with canopy gaps (personal communication).

14. N. J. Collar et al., *The Threatened Birds of the Americas.* 3rd ed., pt. 2 (Washington, D.C.: Smithsonian Institution Press, 1992).

15. A species as rare as the Bachman's warbler is often incompletely characterized, particularly with regard to ecological details. The descriptions here are consistent with historical accounts of the species and with conjectures initially discussed by Paul B. Hamel (personal communication). Hamel's doctoral work involved a systematic but unsuccessful search for Bachman's warblers over large areas of potential habitat in South Carolina. He also summarized historical accounts of the bird's breeding habitat and related this information to the habitats of all the other breeding warblers of the region. See R. G. Hooper and P. B. Hamel, Nesting habitat of Bachman's warbler—a review, unpublished, USDA/Forest Service, Southeastern Forest Experiment Station, Clemson, S.C.; P. B. Hamel and S. A. Gauthreaux, The field identification of Bachman's warbler, *American Birds* 36 (1982):235–240.

16. This example, probably more than any other provided in this book, is speculative. For a creature as consistently rare as the Bachman's warbler, the precise details of the species' true habitat requirements and ecology probably will never be known.

17. H. H. Shugart, *A Theory of Forest Dynamics: The Ecological Implications of Forest Succession Models* (New York: Springer-Verlag, 1984); S. T. A. Pickett and P. S. White, eds., *The Ecology of Natural Disturbance and Patch Dynamics* (New York: Academic Press, 1985); D. C. Glenn-Lewin, R. K. Peet, and T. T. Veblen, eds., *Plant Succession: Theory and Prediction* (London: Chapman and Hall, 1992); H. H. Shugart and T. M. Smith, A review of forest patch models and their application to global change research, *Climatic Change* 34 (1996):131–153; H. H. Shugart, W. R. Emanuel, and G. Shao, Models of forest structure for conditions of climatic change, *Commonwealth Forestry Review* 75 (1996):51–64.

18. Karieva and Anderson found that 95 percent of the studies that they surveyed from leading ecological journals were conducted on plots of less than 1 hectare. Half used plots of 1 square meter or smaller. P. Karieva and M. Anderson, Spatial as-

pects of species interactions: The wedding of models and experiments (pp. 35–50), in A. Hastings, ed., *Community Ecology* (New York: Springer-Verlag, 1988).

19. I have substituted "predictable landscape" for the term "quasi-equilibrium landscape" and "unpredictable landscape" for "effectively nonequilibrium landscape," as used in other books. Shugart, *Forest Dynamics;* H. H. Shugart, *Terrestrial Ecosystems in Changing Environments* (Cambridge: Cambridge University Press, 1998). The average behavior of quasi-equilibrium landscapes is reasonably predictable, hence the name.

20. This section is abstracted from Lindenmayer's very readable account of the life history and issues attending the Leadbeater's possum: D. B. Lindenmayer, *Wildlife and Woodchips: Leadbeater's Possum—A Test Case for Sustainable Forestry* (Sydney: University of New South Wales Press, 1996). Detailed analyses of the possum's habitat are given in D. B. Lindenmayer et al., The conservation of arboreal marsupials in the montane ash forests of the Central Highlands of Victoria, southeast Australia: I, Factors influencing the occupancy of trees with hollows, *Biological Conservation* 54 (1990):111–131; II, The loss of trees with hollows and its implications for the conservation of Leadbeater's possum *Gymnobelides leadbeateri* McCoy (Marsupialia: Petauridae), *Biological Conservation* 54 (1990):133–145; III, The habitat requirements of Leadbeater's possum *Gymnobelides leadbeateri* and models of the diversity and abundance of arboreal marsupials, *Biological Conservation* 56 (1991):295–315.

21. D. H. Ashton, Fire in tall open forests (wet sclerophyll forests) (pp. 339–366), in A. M. Gill, R. H. Groves, and I. R. Noble, eds., *Fire and the Australian Biota* (Canberra: Australian Academy of Science, 1981).

22. A. M. Gill, Post-settlement fire history in Victorian landscapes (pp. 77–98), in Gill, Groves, and Noble, *Fire and the Australian Biota.*

23. N. P. Cheney, Fire behaviour (pp. 151–175), in Gill, Groves, and Noble, *Fire and the Australian Biota.*

24. T. Foster, *Bushfire: History, Prevention and Control* (Sydney: A. H. & A. W. Reed, 1976).

25. D. B. Lindenmayer, Characteristics of hollow-bearing trees occupied by arboreal marsupials in the montane ash forests of the Central Highlands of Victoria, southeast Australia, *Forest Ecology and Management* 40 (1991):289–308.

26. D. B. Lindenmayer and R. C. Lacy, Metapopulation viability of arboreal marsupials in fragmented old-growth forests: Comparisons among species, *Ecological Applications* 5 (1995):183–199.

27. Shugart, *Forest Dynamics;* idem, *Terrestrial Ecosystems.*

28. In this section, a ratio of 1 to 50 for disturbance size to landscape size is used as the boundary between predictable and unpredictable landscapes.

29. Shugart, *Forest Dynamics;* idem, *Terrestrial Ecosystems.*

30. These changes also increase the degree of synchronization of disturbances, which

alters the predictability of landscapes in much the same way as increasing the size of the disturbances.

31. S. R. Kessell, *Gradient Modeling: Resource and Fire Management* (New York: Springer-Verlag, 1979).

32. See R. M. May, *Stability and Complexity in Model Ecosystems* (Princeton: Princeton University Press, 1973).

33. This particular model is for Tasmanian vegetation, which is quite similar to that in Victoria. I. R. Noble and R. O. Slatyer, The use of vital attributes to predict successional changes in plant communities subject to recurrent disturbances, *Vegetatio* 43 (1980):5–21.

34. The details of the various simulations are given in S. W. Seagle and H. H. Shugart, Landscape dynamics and the species-area curve, *Journal of Biogeography* 12 (1985):499–508; Shugart, *Terrestrial Ecosystems.*

35. This concept is not new to ecology. A classic 1962 paper pointed out that smaller wildlife preserves, because of their limited areas and isolation, would eventually lose species. F. W. Preston, The commonness and rarity of species, *Ecology* 48 (1962):254–283.

Chapter 6. The Most Common Bird on Earth

1. M. M. Jaeger et al., Evidence of itinerant breeding in the red-billed quelea, *Quelea quelea,* in the Ethiopian Rift Valley, *Ibis* 128 (1986):469–482.

2. P. Ward, The migration patterns of *Quelea quelea* in Africa, *Ibis* 13 (1971):275–297; J. A. Wiens and M. I. Dyer, Assessing the potential impact of granivorous birds in ecosystems (pp. 205–266), in J. Pinowski and S. C. Kendeigh, eds., *Granivorous Birds in Ecosystems* (Cambridge: Cambridge University Press, 1977).

3. Ibid.

4. P. Ward, Feeding ecology of the black-faced dioch, *Quelea quelea,* in Nigeria, *Ibis* 107 (1965):173–214; idem, The breeding biology of the black-faced dioch, *Quelea quelea,* in Nigeria, *Ibis* 107 (1965):326–349.

5. C. C. H. Elliott, The harvest time method as a means of avoiding quelea damage to irrigated rice in Chad/Cameroon, *Journal of Applied Ecology* 16 (1979):23–35.

6. P. Ward, *Manual of Techniques Used in Research on Quelea Birds* (Rome: UNDP/FAO, 1973); C. C. H. Elliott, The pest status of the quelea (pp. 17–34), in R. L. Bruggers and C. C. H. Elliott, eds., *Quelea quelea, Africa's Bird Pest* (New York: Oxford University Press, 1989).

7. W. C. Mullie, Traditional capture of red-billed quelea, *Quelea quelea,* in the Lake Chad basin and its possible role in reducing damage level in cereals, *Ostrich* 71 (2000):15–20.

8. J. P. Grime, Vegetation classification by reference to strategy, *Nature* 250 (1974):

26–31; idem, Evidence for the existence of three primary strategies in plants and its relevance to ecological and evolutionary theory, *American Naturalist* 111 (1977):1169–94; idem, *Plant Strategies and Vegetation Processes* (Chichester, England: John Wiley, 1979).

9. Disturbance has already been discussed as an agent that changes the pattern of spatial heterogeneity in landscapes (Chapter 4).

10. R. A. A. Oldeman and J. van Dijk, Diagnosis of the temperament of tropical rain forest trees (pp. 21–66), in A. Gómez-Pompa, T. C. Whitmore, and M. Hadley, eds., *Rain Forest Regeneration and Management,* vol. 6 (Paris: UNESCO, 1991); C. G. G. J. van Steenis, Rejuvenation as a factor for judging the status of vegetation types: The biological nomad theory (pp. 212–215), in *Study of Tropical Vegetation,* Proceedings of the Kandy Symposium (Paris: UNESCO, 1958).

11. J. Baird, Returning to the tropics: The epic autumn flight of the blackpoll warbler (pp. 51–62), in K. P. Able, ed., *Gatherings of Angels: Migrating Birds and Their Ecology* (Ithaca, N.Y.: Comstock Books, 1999); B. G. Murray, Jr., A critical review of the transoceanic migration of the blackpoll warbler, *Auk* 106 (1989):8–107. Murray argues for an alternative fall migration route for the blackpoll warbler that involves departure from a region somewhat farther south (between Cape Hatteras and northern Florida).

12. Much of this story has been pieced together in a six-year study by a team of ornithologists using ships at sea and a radar network with sites in Halifax, Nova Scotia; Cape Cod, Massachusetts; Bermuda; Wallops Island, Virginia; Miami, Florida; Antigua; Barbados; and Jamaica. See T. C. Williams et al., Autumnal bird migration over the western North Atlantic Ocean, *American Birds* 31 (1977):251–267; idem, Estimated flight times for North Atlantic migrants, *American Birds* 32 (1978):275–280. Birds and indications of the density of migrants aloft can be detected using radar weather stations.

13. The section that follows is based on K. P. Able, The scope and evolution of migration (pp. 1–11), in Able, *Gatherings of Angels.*

14. Timing cues are known as *Zeitgebers* (German for "time givers") in the standard literature of avian physiology.

15. H. G. Wallraff, Does pigeon homing depend on stimuli perceived during displacement? I, Experiments in Germany, Journal of Comparative Physiology A139 (1980): 193–201.

16. Able, *Gatherings of Angels.*

17. R. Wiltschko and W. Wiltschko, *Magnetic Orientation in Animals* (Berlin: Springer-Verlag, 1995).

18. This remarkable capability is described in K. P. Able, A sense of magnetism, *Birding* 30 (1998):314–321.

19. Idem, How birds migrate (pp. 11–26), in Able, *Gatherings of Angels.*

20. K. Schmidt-Koenig and H.-J. Schlichte, Homing in pigeons with reduced vision, *Proceedings of the National Academy of Sciences USA* 69 (1972):2446–47.

21. M. M. Walker et al., Structure and function of the vertebrae magnetic sense, *Nature* 390 (1997):371–376.

22. F. Papi, Pigeons use olfactory clues to navigate, *Ethological and Ecological Evolution* 1 (1989):219–231; K. P. Able, The debate over olfactory homing in pigeons, *Journal of Experimental Biology* 199 (1996):121–124.

23. W. L. Donn and B. Naini, The sea-wave origin of microbaroms and microseisms, *Journal of Geophysical Research* 78 (1973):4428–88; J. T. Hagstrum, Infrasound and the avian navigational map, *Journal of Experimental Biology* 203 (2000):1103–11.

24. P. Berthold, *Control of Bird Migration* (London: Chapman and Hall, 1996).

25. Able, *Gatherings of Angels.*

26. R. T. Peterson, How many birds are there? *Audubon Magazine* 43 (1941):142. Peterson estimated the population of the forty-eight contiguous United States to be over 5.5 billion in the summer breeding season and close to 4 billion in the winter season.

27. J. Cartier, *Voyage de Jacques Cartier en 1534,* ed. M. H. Michelant.

28. Schorger uses the low range of historical accounts of the sizes of species flocks to obtain this estimate. Some of the less conservative estimates would produce a still higher proportion.

29. Some observers have reported as much as a half a meter of bird droppings. Investigations into possible use of the guano to produce gunpowder components estimated the amounts in wagonloads.

30. J. J. Audubon, *Ornithological Biography* 1 (1841):323.

31. A. W. Schorger, *The Passenger Pigeon: Its Natural History and Extinction* (Norman: University of Oklahoma Press, 1973). There are recorded instances of single nests or nests of a few birds, but extremely large nestings were the norm for the species.

32. Ibid.

33. The smaller nesting groups appear to have been mostly young birds and may represent adherent behavior in immature individuals.

34. Table 2 (p. 57) in Schorger, *The Passenger Pigeon,* summarizes information from a number of sources.

35. W. Byrd, *William Byrd's Histories of the Dividing Line betwixt Virginia and North Carolina,* 1728; later ed. by W. K. Boyd (Raleigh: North Carolina Historical Commission, 1929). Quotation on p. 216.

36. The demise of the passenger pigeon appears to be a direct consequence of human actions. Nonetheless, it has been argued that an introduced disease of European pigeons could have caused the decline. Great numbers of birds were reported to have drowned in storms encountered by flocks traversing extensive water bodies. Pigeon roosts and nestings were worked over by several bird and mammal preda-

tors, but the flocking behavior is generally thought to have reduced the overall impact of predation.

37. Recall the second collection of Bachman's warblers related in Chapter 5.

Chapter 7. The Engineering Rodent

1. C. Maser et al., *Natural History of Oregon Coast Mammals* (Portland, Ore.: USDA/ Forest Service, 1981).

2. E. T. Seton, *Lives of Game Animals* (Garden City, N.Y.: Doubleday, Page, 1929).

3. G. C. Cline, *Peter Skene Odgen and the Hudson's Bay Company* (Norman: University of Oklahoma Press, 1974).

4. D. R. Johnson and D. H. Chance, Presettlement overharvest of upper Columbia River beaver populations, *Canadian Journal of Zoology* 52 (1974):1519–21; S. H. Jenkins and P. E. Busher, *Castor canadensis, Mammal Species* 120 (1979):1–9.

5. R. J. Naiman, C. A. Johnston, and J. C. Kelly, Alteration of North American streams by beaver, *BioScience* 38 (1988):753–762.

6. See M. Kurlanski, *Cod: A Biography of the Fish That Changed the World* (New York: Walker, 1997).

7. R. Rudemann and W. J. Schoonmaker, Beaver dams as geological agents, *Science* 88 (1938):523–525.

8. Kurlanski, *Cod.*

9. J. E. DeKay, *Zoology of New York,* Natural History of New York, 1842 (available from the New York Public Library's Research Library).

10. Kurlanski, *Cod.*

11. Ibid.

12. Dugmore, *Romance of the Beaver.*

13. Ibid.

14. Naiman, Johnston, and Kelly, Alteration of North American streams; R. J. Naiman et al., Beaver influences on the long-term biogeochemical characteristics of boreal forest drainage networks, *Ecology* 75 (1994):905–921.

15. W. Bartram, *The Travels of William Bartram,* 1791 (New York: reprint ed. Dover Publications, 1928); J. E. Bakeless, *The Eyes of Discovery* (New York: Dover Publications, 1961); J. F. Triska, Role of woody debris in modifying channel geomorphology and riparian areas of a large lowland river under pristine conditions: An historical case study, *Internationale Vereinigung für theoretische und angewandte Limnologie, Vehandlungen* 22 (1984):1828–92.

16. R. I. E. Newell, Ecological changes in Chesapeake Bay: Are they the results of overharvesting the American oyster, *Crassostrea virginica?* (pp. 536–546), in *Understanding the Estuary: Advances in Chesapeake Bay Research,* Proceedings, Chesapeake Research Consortium, 1988; M. G. McCormick-Ray, Oyster reefs in 1878 seascape pattern—Winslow revisited, *Estuaries* 21 (1998):784–800.

17. G. S. Brush, Rates and patterns of estuarine sediment accumulation, *Limnology and Oceanography* 34 (1989):1235–46.

18. Newell, Ecological changes; R. E. Ulanowicz and J. H. Tuttle, The trophic consequences of oyster stock rehabilitation in Chesapeake Bay, *Estuaries* 15 (1992):298–306; McCormick-Ray, Oyster reefs.

19. R. Rudemann and W. J. Schoonmaker, Beaver dams as geological agents, *Science* 88 (1938):523–525; R. L. Ives, The beaver-meadow complex, *Journal of Geomorphology* 5 (1942):191–203.

20. C. S. Darwin, *The Formation of Vegetable Mould through the Action of Worms with Observations on Their Habits* (New York: Humboldt, 1887).

21. P. Coxon and S. Waldren, Flora of the Quaternary temperate stages of NW Europe: Evidence for large-scale changes (pp. 104–117), in B. Huntley et al., *Past and Future Rapid Environmental Changes: The Spatial and Evolutionary Responses of Terrestrial Biota* (Berlin: Springer-Verlag, 1997).

22. In fact, the African systems have been referred to as the "odd man out" among the tropical rain forests of the world. B. J. Meggers, E. S. Ayensu, and W. D. Duckworth, eds., *Tropical Forest Ecosystems in Africa and South America: A Comparative Review* (Washington, D.C.: Smithsonian Institution Press, 1973).

23. P. J. Regal, Ecology and evolution of flowering plant dominance, *Science* 196 (1977):622–629.

24. This adaptation is reviewed in L. S. Adler, The ecological significance of toxic nectar, *Oikos* 91 (2000):409–420.

25. G. J. Keighery, Bird-pollinated plants in Western Australia (pp. 77–89), in J. A. Armstrong, J. M. Powell, and A. J. Richards, eds., *Pollination and Evolution* (Sydney: Royal Botanic Gardens, 1982).

26. R. J. Lambeck, The role of faunal diversity in ecosystem function (pp.129–148), in R. J. Hobbs, ed., *Biodiversity in Mediterranean Ecosystems in Australia* (Norton, New South Wales: Surrey Beatty, 1992).

27. N. E. Walker, *Soil Microbiology* (London: Butterworths, 1957).

Chapter 8. The Fall of the Big Bird

1. By comparison, the ostrich (*Struthio camelus*) is the tallest living bird at 8 feet (2.5 m).

2. M. M. Trotter and B. McCulloch, Moas, men and middens (pp. 709–728), in P. S. Martin and R. G. Klein, eds., *Quaternary Extinctions: A Prehistoric Revolution* (Tucson: University of Arizona Press, 1984).

3. B. Gill and P. Martinson, *New Zealand's Extinct Birds* (Auckland: Random Century, 1991).

4. R. Duff, *The Moa-Hunter Period of Maori Culture* (Wellington: E. C. Keating [Government Printer], 1950).

5. C. J. Burrows, Diet of the New Zealand Dinornithoformes, *Naturwissenschafften* 67 (1980):S151; Trotter and McCulloch, Moas, men and middens.

6. Duff, in *The Moa-Hunter Period,* reports moa remains on *top* of tussock grass that dates to the year 230 (60 B.P.) using the ^{14}C-isotope dating technique. Other possible post-European records are discussed in A. J. Anderson, The extinction of moa in southern New Zealand (pp. 728–740), in Martin and Klein, *Quaternary Extinctions;* and Trotter and McCulloch, Moas, men and middens. Scientific debate persists regarding these records. It seems that the moa came tantalizingly close to surviving until European contact. The existence of any moas today is highly unlikely.

7. Anderson, Extinction of moa.

8. Ibid.

9. M. S. McGlone, Polynesians and the late Holocene deforestation of New Zealand (p. 43), in A. Ross, ed., *Environment and People in Australasia,* abstracts of the 52nd Congress of the Australian and New Zealand Association for the Advancement of Science, Macquarie University, Sydney, Australia.

10. The location was Shag Mouth on the South Island. See H. D. Skinner, Archeology of Canterbury: II, Monck's Cave, *Records of the Canterbury Museum* 2 (1924):151–162. Trotter and McCulloch, Moas, men and middens, cite several other examples.

11. A. J. Anderson, A review of economic patterns during the Archaic phase in southern New Zealand, *New Zealand Journal of Archeology* 3 (1982):15–20.

12. E. Skokstad, Divining diet and disease from DNA, *Science* 289 (2000):530–531.

13. Anderson, in The extinction of moa, produced a "back-of-the-envelope" but conservative estimate of 100,000 to 500,000 moas in the 117 known moa-hunting sites. He based his numbers on the size of the sites and the density of moa bones found. He further noted that the largest sites are on a coastline that has been receding at the rate of 0.5 to 1 m per year, so some major butchering sites may not be included in his calculations. Anderson feels that the size of the moa population was probably in the tens of thousands. He later noted that the count on sites has risen to 300 and reiterated the role of overhunting in eliminating these species. A. J. Anderson, Prehistoric Polynesian impact on the New Zealand environment: Te Whenua Hou (pp. 271–283), in P. V. Kirch and T. L. Hunt, eds., *Historical Ecology in the Pacific Islands* (New Haven: Yale University Press, 1995). Also see idem, Mechanics of overkill in the extinction of New Zealand moas, *Journal of Archeological Science* 16 (1989):137–151; and idem, *Prodigious Birds: Moas and Moa-Hunting in Prehistoric New Zealand* (Cambridge: Cambridge University Press, 1989).

14. Technological human societies have demonstrated this fact by virtually eliminating the great whales of the oceans in about half the time postulated for the moa extinctions.

15. R. N. Holdaway and C. Jacomb, Rapid extinction of the moas (Aves: Diornithiformes): Model, test and implications, *Science* 287 (2000):2250–54.

16. Gill and Martinson, *New Zealand's Extinct Birds;* Anderson, in Prehistoric Polyne-

sian impact, puts the number of extinctions at thirty-seven species *and subspecies.* The inclusion of subspecies in the count would of course tend to produce a larger total.

17. Anderson, Prehistoric Polynesian impact.

18. R. Cassels, The role of prehistoric man in the faunal extinctions of New Zealand and other Pacific islands (pp.741–767), in Martin and Klein, *Quaternary Extinctions.*

19. S. L. Olson and H. F. James, *Descriptions of Thirty-Two New Species of Birds from the Hawaiian Islands,* pt. 1, *Non-Passeriformes,* Ornithological Monograph no. 45, American Ornithologists Union, Washington, D.C., 1991; H. F. James and S. L. Olson, *Descriptions of Thirty-Two New Species of Birds from the Hawaiian Islands,* pt. 2, *Passeriformes,* Ornithological Monograph no. 46, American Ornithologists Union, Washington, D.C., 1991. The two ornithologists have also developed an exceptional body of work on the fossil and recent birds of other Pacific islands.

20. Among them are several large flightless geese, five flightless rails, and several species of owls with long legs and short wings that were predators on small birds, and sixteen new and extinct species of songbirds (Passerines).

21. Steadman puts the number of extinct fossil birds at sixty species known only from bones, plus another twenty to twenty-five species that have become extinct in the last 200 years. D. W. Steadman, Prehistoric extinctions of Pacific Island birds: Biodiversity meets zooarcheology, *Science* 267 (1995):1123–31.

22. S. L. Olson and H. F. James, The role of Polynesians in the extinction of the avifauna of the Hawaiian Islands (pp. 768–780), in Martin and Klein, *Quaternary Extinctions;* idem, *Thirty-Two New Species,* pt. 1, quotation on p. 7; James and Olson, *Thirty-Two New Species,* pt. 2.

23. This evocative term is credited to A. W. Crosby, *Ecological Imperialism: The Biological Expansion of Europe, 900–1900* (Cambridge: Cambridge University Press, 1986).

24. J. S. Athens, Hawaiian native lowland vegetation in prehistory (pp. 248–270), in Kirch and Hunt, *Historical Ecology.*

25. Many of these plants would not have survived the frosts of temperate New Zealand, a fact that is likely to have reinforced Maori dependency on the native flora and fauna.

26. P. A. Cox and S. A. Banack, eds., *Islands, Plants and Polynesians: An Introduction to Polynesian Ethnobotany* (Portland, Ore.: Dioscorides Press, 1991). Several chapters outline both the use and the transportation of plants by Polynesians and are a useful summary of this topic.

27. P. V. Kirch, Changing landscapes and sociopolitical evolution in Mangaia, Central Polynesia (pp. 147–165), in Kirch and Hunt, *Historical Ecology.*

28. Ibid.

29. Consider the following as an example: The first man who, having enclosed a piece

of ground, thought of saying "This is mine," and found people simple enough to believe him, was the true founder of civil society. Humanity would have been spared infinite crimes, wars, homicides, and murders if only someone had ripped up the fences or filled in the ditches and said: "Do not listen to this pretender! You are eternally lost if you do not remember that the fruits of the earth are everyone's property and that the land is no-one's property!" From Jean-Jacques Rousseau, *The Social Contract and Discourses,* 1754. Translated by G. D. Cole (London: J. M. Dent, 1913), p. 207.

30. P. V. Kirch, Changing landscapes and sociopolitical evolution in Mangaia, Central Polynesia (pp. 147–165), in Kirch and Hunt, *Historical Ecology.* Quotation on p. 4.

31. Trotter and McCulloch, Moas, men and middens. Quotation on p. 723.

32. Gill and Martinson, *New Zealand's Extinct Birds.* These records are not strictly comparable, since the determination of European extinctions includes a written historical record that is lacking in the Maori record. As will be discussed in the chapters that follow, Europeans have produced remarkable levels of extinctions across the planet.

33. The species appears to have been widespread on mainland New Zealand prior to Polynesian colonization and may have been eliminated across most of the country by the introduced Polynesian rat. R. N. Holdaway, New Zealand's pre-human avifauna and its vulnerability, in M. R. Rudge, ed., Moas, man and climate in the ecological history of New Zealand, *New Zealand Journal of Ecology* 12 (suppl.) (1989):11–25.

34. Martin and Klein, *Quaternary Extinctions.*

35. G. Singh, A. P. Kershaw, and R. Clark, Quaternary vegetation and fire history in Australia (pp. 23–54), in A. M. Gill, R. H. Groves, and I. R. Noble, eds., *Fire and the Australian Biota* (Canberra: Australian Academy of Sciences, 1981); P. H. Nicholson, Fire and the Australian aborigine—An enigma (pp. 55–76), in Gill, Groves, and Noble, *Fire and the Australian Biota.*

36. G. Barker, *Prehistoric Farming in Europe* (Cambridge: Cambridge University Press, 1985).

37. Archaic Indians: P. A. Delcourt, Goshen Springs: Late-Quaternary vegetation record for southern Alabama, *Ecology* 61 (1980):371–386; Australia: Nicholson, Fire and the Australian aborigine.

38. Steadman, Prehistoric extinctions.

39. E. G. Munroe, The size of island faunas (pp. 52–53), in *Proceedings of the Seventh Science Congress of the Pacific,* vol. 4, *Zoology* (Auckland: Whitcome and Tombs, 1953); R. H. MacArthur and E. O. Wilson, An equilibrium theory of insular zoogeography, *Evolution* 17 (1963):373–387; idem, *The Theory of Island Biogeography* (Princeton: Princeton University Press, 1967).

40. Ibid.

41. This time interval should be the inverse of the turnover rate and is referred to in other contexts as the turnover time.

42. A readable review and summarization of the diversity of ideas on island biogeography is in R. J. Whitaker, *Island Biogeography* (Oxford: Oxford University Press, 1998).

43. M. B. Bush and R. J. Whitaker, Non-equilibration in island theory of Krakatau, *Journal of Biogeography* 20 (1993):453–456; Whitaker, *Island Biogeography.*

44. The islands are called the Krakatau group after the name of the largest of the three islands.

45. While the colonization of Rakata is a massive "natural experiment" in the colonization and extinction processes of islands, it is far from a perfectly designed experiment. The island is difficult to survey. In 1927 a new volcano called Anak Karkatua started in the middle of the caldera created by the destruction of Krakatau. This volcanic island is active and periodically disturbs the surrounding islands, particularly affecting the forests.

46. Initial observations in 1884 from R. D. M. Verbeek, *Krakatau* (in French), pts. 1 and 2, pp. 396–461 (Batavia: Government Printing Office, 1886); see Whitaker, *Island Biogeography,* and elsewhere.

47. These were *Thalassochorous* species in one classification scheme of plant dispersal mechanisms. L. van der Pijl, *Principles of Dispersal in Higher Plants* (Berlin: Springer-Verlag, 1972).

48. *Amenochorous* species (ibid.).

49. *Zoochorous* species (ibid.).

50. Whitaker, *Island Biogeography.*

51. The warming of the average temperature of the ocean causes the waters to expand. This "thermal expansion" also causes the sea level to rise—even in the absence of additional water from the melting of glaciers on the terrestrial surface.

52. J. Weins, On understanding a non-equilibrium world: Myth and reality in community patterns and processes (pp. 439–457), in D. R. Strong, Jr., et al., eds., *Ecological Communities: Conceptual Issues and the Evidence* (Princeton: Princeton University Press, 1984). Weins points out that it is inappropriate to assume that a system is at equilibrium without actually demonstrating it to be so. Most observations of natural systems with any degree of duration illustrate a tendency for change, not the constancy implied by an equilibrium condition. Pregill and Olson point to the need to better understand both historical data and the effects of environmental change. G. K. Pregill and S. L. Olson, Zoogeography of West Indian vertebrates in relation to Pleistocene climate cycles, *Annual Review of Ecology and Systematics* 12 (1981):75–98.

53. J. D. Sauer, Oceanic islands and biogeographical theory: A review, *Geographical Review* 59 (1969):582–593.

54. Data on species turnover are difficult to collect. See J. D. Lynch and N. V. Johnson,

Turnover and equilibria in insular avifaunas, with special reference to the California Channel Islands, *Condor* 76 (1974):370–384. The data are confounded by interactions between trophic levels. See G. L. Hunt, Jr., and M. W. Hunt, Trophic levels and turnover rates: The avifauna of Santa Barbara Island, *Condor* 76 (1974):363–369. The turnover of species may involve transient species that are not really colonists but simply alight on the island and then move elsewhere. See D. Simberloff, Species turnover and equilibrium island biogeography, *Science* 194 (1976):572–578.

55. M. H. Williamson, *Island Populations* (Oxford: Oxford University Press, 1981); idem, The MacArthur and Wilson theory today: True but trivial, *Journal of Biogeography* 16 (1989):3–4; idem, Natural extinction on islands, *Philosophical Transactions of the Royal Society of London, Series B*, 325 (1989):457–468.

56. F. S. Gilbert, The equilibrium theory of island biogeography, fact or fiction? *Journal of Biogeography* 7 (1980):209–235.

57. Simberloff, Species turnover.

58. J. B. Foster, Evolution of mammals on islands, *Nature* 202 (1965):234–235; M. V. Lomolino, Body size of mammals on islands: The island rule reexamined, *American Naturalist* 125 (1985):310–316.

59. R. A. Reyment, Palaeontological aspects of island biogeography: Colonization and evolution of mammals on Mediterranean islands, *Oikos* 41 (1893):299–306.

60. B. Groombridge, ed., *Global Biodiversity: Status of the Earth's Living Resources* (London: Chapman and Hall, 1992); D. W. Steadman, Human-caused extinctions of birds (pp. 139–161), in M. L. Reaka-Kudla, W. E. Wilson, and W. O. Wilson, eds., *Biodiversity II: Understanding and Protecting Our Natural Resources* (Washington, D.C.: Joseph Henry Press, 1997).

61. Whitaker, *Island Biogeography.*

Chapter 9. The Wolf That Was Woman's Best Friend

1. Bear cubs in North America: F. Galton, The first steps toward the domestication of animals, *Transactions of the Ethnological Society of London, N.S.* 3 (1865):122–138); in China: J. G. Frazer, *The Golden Bough: A Study in Magic and Religion* (London: Macmillan, 1922). Monkeys, kinkajous and other creatures in South America: W. E. Roth, An introductory study of the arts, crafts, and customs of the Guiana Indians (pp. 25–745), in F. W. Hodge (transmitter), accompanying paper to the thirty-eighth *Annual Report of the Bureau of American Ethnology to the Secretary of the Smithsonian Institution, 1916–1917* (Washington, D.C.: Government Printing Office, 1924). J. Serpell, Pet-Keeping and animal domestication: A reappraisal (pp. 10–21), in J. Clutton-Brock, ed., *The Walking Larder* (London: Unwin Hyman, 1989), reviews the role of women in taming a variety of animals for pets in a wide range of cultures. Australian Aboriginals: G. Krefft, *The Mammals of Aus-*

tralia (Sydney: Printer, 1871). New Guineans: M. Titcomb, *Dog and Man in the Ancient Pacific* (Honolulu: Bernice P. Bishop Museum, 1969).

2. K. Dennis-Bryan and J. Clutton-Brock, *Dogs of the Last Hundred Years at the British Museum (Natural History)* (London: British Museum [Natural History], 1988).

3. A. P. Gray, *Animal Hybrids: A Checklist with Bibliography* (Bucks, UK: Farnham Royal, 1954).

4. C. Vilà et al., Man and his dog, *Science* 278 (1997):206–207; idem, Multiple and ancient origins of the domestic dog, *Science* 276 (1997):1687–89.

5. F. E. Zeuner, *A History of Domesticated Animals* (London: Hutchinson, 1963).

6. J. K. Gollan, Prehistoric dingo, doctoral thesis, Australian National University, 1982.

7. The earlier records (ca. 8500 B.P.) of dingoes in Australia are for isolated teeth, which may have dropped from an earlier level during excavation of the archeological site. S. J. Olsen, *Origins of the Domesticated Dog: The Fossil Record* (Tuscon: University of Arizona Press, 1985), believes that the evidence is too meager for a definitive determination; subsequent scientists have largely upheld this opinion.

8. L. K. Corbett, Morphological comparisons of Australian and Thai dingoes: A reappraisal of dingo status, distribution and ancestry, *Proceedings of the Ecological Society of Australia* 13 (1985):277–291, favors the Thai dog as a source for the dingo; Gollan, Prehistoric dingo, leans to the Indian pariah dog.

9. Corbett, Morphological comparisons.

10. Idem, *The Dingo in Australia and Asia* (Sydney: University of New South Wales Press, 1995).

11. M. Titcomb, *Dog and Man in the Ancient Pacific* (Honolulu: Bernice P. Bishop Museum, 1969).

12. A. Newsome (personal communication), who conducted CSIRO studies on the biology and genetics of dingoes, relates that dingo pups taken into captivity after their eyes opened were always difficult to handle. None ever became tame enough to answer to commands.

13. R. A. Gould, Journey to Pulykara, *Natural History* 79 (1970):57–66.

14. M. W. Fox, *The Wild Canids* (New York: Van Nostrand Reinhold, 1975); Olsen, *Origins of the Domesticated Dog.*

15. Corbett, *The Dingo in Australia and Asia.*

16. Margaret W. Smith, personal communication.

17. P. Savolainen et al., Genetic evidence for an East Asian origin of domestic dogs, *Science* 298 (2002):1610–13.

18. References and other documentation for a variety of sites can be found in Olsen, *Origins of the Domesticated Dog.* D. F. Morey and M. Wiant, Early Holocene domestic dog burials from the North American Midwest, *Current Anthropology* 33 (1992):224–229, provide records of early dogs from New World archeological sites. Genetic evidence that the dogs of the New World originated from Asian domesticated wolves subsequently brought to the New World can be found in J. A.

Leonard et al., Ancient DNA evidence for Old World origin of New World dogs, *Science* 298 (2002):1613–16.

19. The time scale in Figure 36 is based on the assumption that the wolf-like canids and the South American canids had *Canis davisii* as a common ancestor, as proposed by A. Berta, The Pleistocene bush dog, *Speothus pacivorus* (Canidae), from the Lagoa Santa caves, Brazil, *Journal of Mammalogy* 65 (1984):549–559. The now-extinct *Canis davisii* appeared in the fossil record about 7 million years ago, so that 0.1 genetic distance unit $\cong$ 2.5 million years.

20. Savolainen et al., Genetic evidence; Vilà et al., Man and his dog; idem, Multiple and ancient origins.

21. See J. P. Scott, O. S. Elliot, and B. E. Ginsburg, Man and his dog, *Science* 278 (1997):205; and N. E. Federoff and R. M. Nowak, Man and his dog, *Science* 278 (1997):205, written in response to Vilà et al., Multiple and ancient origins.

22. Olsen, *Origins of the Domesticated Dog.*

23. H. de Lumley, Une cabana de chasseuse acheuléens dans la grotte du Lazaret à Nice, *Archeologia* 28 (1969):26–33. J. Clutton-Brock, *A Natural History of Domesticated Mammals* (Austin: University of Texas Press, 1989), notes that L. R. Binford feels the animal remains are from a wolf den in the cave and unrelated to human ritual.

24. Clutton-Brock, *A Natural History.*

25. A possible exception would be if young tamed animals had been castrated to produce larger and/or more docile individuals. Clutton-Brock mentions this scenario as a possible means of determining the time of domestication of the reindeer. An analysis demonstrating such a situation would require relatively large sample sizes to demonstrate that the animal remains associated with humans were different from those of the local wild population.

26. J. Clutton-Brock. 1969. Carnivore remains from evacuations of the Jericho Tell (pp. 357–345). In: P. Ucko and G. W. Dimblely (ed.). *The Domestication of Plants and Animals.* Aldine Publishing, Chicago. Quotation on p. 340.

27. Olsen, *Origins of the Domesticated Dog.*

28. I. G. Plidochichko, *Late Paleolithic Dwellings of Mammoth Bones in the Ukraine* (Naukova Dumka, Kiev: Institute of Zoology of the Ukrainian Academy of Sciences, 1969).

29. Olsen, *Origins of the Domesticated Dog.*

30. See Clutton-Brock, *A Natural History.*

31. Clutton-Brock developed this list from an equivalent earlier list in Galton, The first steps toward the domestication of animals. *Transactions of the Ethnological Society of London, N.S.* 3 (1865):122–138. The original Galton list is quoted in Chapter 11.

32. I. L. Mason, *Evolution of Domesticated Animals* (London: Longman, 1982).

33. D. E. MacHugh and D. G. Bradley, Livestock genetic origin: Goats buck the trend, *Proceedings of the National Academy of Science, USA* 98 (2001):5382–84.

34. G. Luikart et al., Multiple maternal origins and weak phylographic structure in domestic goats, *Proceedings of the National Academy of Science, USA* 98 (2001): 5927–32.

35. Clutton-Brock, *A Natural History.*

36. G. Luikart et al., Multiple maternal origins.

37. N. J. Wood and S. H. Phua, Variation in the control region sequence of the sheep mitochondrial genome, *Animal Genetics* 27 (1996):25–33.

38. Clutton-Brock, *A Natural History.*

39. E. Giuffra et al., The origin of the domestic pig: Independent domestication and subsequent introgression, *Genetics* 154 (2000):1785–91.

40. Clutton-Brock, *A Natural History.*

41. Ibid.

42. C. Vilà et al., Widespread origins of domestic horse lineages, *Science* 291 (2001): 474–477.

43. G. Luikart et al., Multiple maternal origins.

44. Clutton-Brock, *A Natural History.*

45. Zeuner, *A History of Domesticated Animals.*

46. However, pigs are often reported as being nursed by humans in New Guinea and elsewhere.

47. Onagers have the humorous common name of "half-asses" to connote their intermediate size between horses and asses.

48. Zeuner, *A History of Domesticated Animals.*

49. J. Clutton-Brock, Carnivore remains from evacuations of the Jericho Tell (pp. 337–345), in P. Ucko and G. W. Dimblely, eds., *The Domestication and Exploitation of Plants and Animals* (Chicago: Aldine Publishing, 1969).

50. M. Hopf, Plant remains and early farming in Jericho (pp. 355–359), in Ucko and Dimbleby, *Domestication and Exploitation.*

51. Clutton-Brock, Carnivore remains.

52. Clutton-Brock, *A Natural History.*

53. P. L. Fall, C. A. Lindquist, and S. E. Falconer, Fossil hyrax middens from the Middle East: A record of paleovegetation and human disturbance (pp. 408–427), in J. L. Betancourt, T. R. Van Devender, and P. S. Martin, eds., *Packrat Middens: The Last 40,000 Years of Change* (Tucson: University of Arizona Press, 1990).

54. J. F. Downs, Comments on the plains Indians cultural development, *American Anthropologist* 66 (1964):421–422.

55. E. West, *The Way to the West* (Albuquerque: University of New Mexico Press, 1995). Quotation on p. 22.

56. J. Doring, *Kulturwandel bei den Nordamerikanischen Plainsindianern: Zur Rolle des Pferdes bei den Comanchen und den Cheyenne* (Berlin: Dietrick Reimer, 1984).

57. A. Leopold, *Game Management* (New York: Charles Scribner's Sons, 1933).

58. A. R. Ek and R. A. Monserud, *FOREST: Computer Model for the Growth and Re-*

production Simulation for Mixed Species Forest Stands (Madison: University of Wisconsin, College of Agricultural and Life Sciences, 1974); idem, Trials with program FOREST: Growth and reproduction simulations for mixed species even- or uneven-aged forest stands (pp. 56–73), in J. Fries, ed., *Growth Models for Tree and Stand Simulation* (Stockholm: Royal College of Forestry, 1974); J. W. Ranney, Edges of forest islands: Structure, composition, and importance to regional forest dynamics, doctoral dissertation (Knoxville: University of Tennessee, 1978).

59. For example, mountaintops are isolated ecosystems for many of the species found there, and are closer to being analogous to islands than agricultural landscapes.

60. A sustainable population in this sense would be a population with a high probability of surviving over a relatively long period.

Chapter 10. The Gentle Invader

1. H. V. Thompson, The rabbit in Britain (pp. 62–107), in H. V. Thompson and C. M. King, eds., *The European Rabbit: The History and Biology of Successful Colonizer* (Oxford: Oxford University Press, 1994).

2. G. B. Corbet, Taxonomy and origins (pp. 1 to 7), in Thompson and King, *The European Rabbit*. Reumer and Sanders argue that the Phoenicians brought the species to North Africa. J. W. F. Reumer and E. A. C. Sanders, Changes in the vertebrate fauna of Menorca in prehistoric and classical times, *Zeitschrift für Säugertierkunde* 49 (1984):321–325.

3. C. Lever, *Naturalized Mammals of the World* (London: Longman, 1985).

4. Lagomorph is a general name for species in the mammalian order Lagomorpha, as rodent is for the order Rodentia. Lagomorphs differ from rodents in that the testes are located in front of the penis (as is the case in marsupials). Also, the second set of incisor teeth lie behind the first, not beside them as in rodents and other mammals. J. Clutton-Brock, *A Natural History of Domesticated Mammals* (Austin: University of Texas Press, 1989).

5. J. Sheail, *Rabbits and Their History* (Newton Abbot, England: David and Charles, 1971).

6. Corbet reviewed the taxonomy of rabbits and related hares and found this distinctive behavior the only justification for distinguishing the genus *Oryctolagus* from other rabbits—notably the genus *Silvalagus*, which includes the cottontail rabbits of the New World. G. B. Corbet, A review of the classification in the family Leporidae, *Acta Zoologica Fennica* 174 (1983):11–15.

7. G. E. H. Barrett-Hamilton, *A History of British Mammals*, vol. 2 (London: Gurney and Jackson, 1912).

8. H. Nauchtsheim, *Vom Wildtier zum Haustier* (Berlin: Paul Parey, 1949).

9. J. E. C. Flux, World distribution (pp. 8–21), in Thompson and King, *The European Rabbit*.

10. Clutton-Brock, *A Natural History;* P. M. Rogers, C. P. Arthur, and R. C. Soriguer, The rabbit in continental Europe (pp. 22–63), in Thompson and King, *The European Rabbit.*

11. Flux, World distribution.

12. Thompson, The rabbit in Britain.

13. J. E. C. Flux and P. J. Fullagar, World distribution of the rabbit *Oryctolagus cuniculus* on islands, *Mammal Review* 22 (1992):151–205.

14. For example, the first historical record on the Scilly Islands dates from 1176. Sheail, *Rabbits and Their History.*

15. Thompson, The rabbit in Britain.

16. Andrew Grey, Commissary to the colony, on May 1, 1788, listed five rabbits in An account of livestock in the settlement, Lever, *Naturalized Mammals.*

17. For a historical account of the rabbit in these early years, see Lever, *Naturalized Mammals;* and E. C. Rolls, *They All Ran Wild: The Story of Pests on the Land in Australia* (Sydney: Angus and Robertson, 1969). Quotation on p. 13.

18. A feral species is one that has reverted to the wild state after having been domesticated.

19. Reported in 1869 by James Calder, surveyor-general for Tasmania, Lever, *Naturalized Mammals.*

20. Rolls, *They All Ran Wild.*

21. G. C. Caughley, *Analysis of Vertebrate Populations* (New York: John Wiley, 1977).

22. H. M. Neave, *Rabbit Calicivirus Disease Program Report 1: Overview of the Effects on Australian Wild Rabbit Populations and Implications for Biodiversity,* a report of research conducted by participants in the Rabbit Calicivirus Disease Monitoring and Surveillance Program and Epidemiological Research Program, prepared for the Bureau of Rural Sciences, Canberra.

23. R. M. MacDowall, Exotic fishes: The New Zealand experience (pp. 200–214), in W. R. Courtenay, ed., *Distribution, Biology, and Management of Exotic Fishes* (Baltimore: Johns Hopkins University Press, 1984).

24. F. C. Kinsky, Amendments and additions to the 1970 annotated checklist of the birds of New Zealand, *Notornis* 27 (suppl.) (1980):1–3.

25. I. A. E. Atkinson and E. K. Cameron, Human influence on the terrestrial biota and biotic communities of New Zealand, *Trends in Ecology and Evolution* 8 (1993):447–451.

26. Steadman estimates that two thousand species of birds have been eliminated from the tropical Pacific by prehistoric human settlement. Over half of these were flightless rails, which are found in archeological remains on a number of islands and are presumed to have been endemic species. D. W. Steadman, Prehistoric extinctions of Pacific island birds: Biodiversity meets zooarcheology, *Science* 267 (1995):1123–31).

27. H. W. Simmonds, *My Weapons Had Wings: The Adventures of a Government En-*

tomologist Based in Fiji for Forty-Five Years (Auckland: Percy Salman, Wills and Grainger, 1964).

28. The initial Fijian range of the species was in the vicinity of two small islets, Bau and Viwa, which were used by sandalwood traders.

29. Simmonds' autobiography contains a marvelous collection of stories told from the point of view of a colonial British expatriate traveling the South Pacific. This section is derived from his account, *My Weapons Had Wings.*

30. V. Vincek et al., How large was the founding population of Darwin's finches? *Proceedings of the Royal Society of London B* 264 (1997):111–118.

31. R. J. Whitaker, *Island Biogeography* (Oxford: Oxford University Press, 1998).

32. Q. C. B. Cronk and J. L. Fuller, *Plant Invaders* (London: Chapman and Hall, 1995).

33. L. L. Manne, T. M. Brooks, and S. L. Pimm, Relative risk of extinction of passerine birds on continents and islands, *Nature* 399 (1999):258–261.

34. P. M. Vitousek et al., Introduced species: A significant component of human-caused global change, *New Zealand Journal of Ecology* 21 (1997):1–16.

35. The disease in rabbits caused by the virus is called myxomatosis. The virus, discovered in Uruguay in 1896, was used to drastically reduce the European rabbit population in Australia, with initial releases in 1950.

36. P. F. A. Delille, Une methode nouvelle permettant à l'agriculture de lutter efficacement contre la pullulation du lapin, *Contes Rendus de l'Académie Agricultural Français* 39 (1953):638.

37. This account is from P. M. Rogers, C. P. Arthur, and R. C. Soriguer, The rabbit in continental Europe (pp. 22–63), in Thompson and King, *The European Rabbit.* The sums are based on 1992 values of the franc.

38. F. Fenner and F. N. Ratcliffe, *Myxomatosis* (Cambridge: Cambridge University Press, 1965).

39. Aragão's proposal was made through Dr. A. Breinl, then director of the Australian Institute of Tropical Medicine in Townsville, Queensland.

40. Fenner and Ratcliffe, *Myxomatosis.*

41. C. K. Williams, Ecological challenges to controlling wild rabbits in Australia using virally-vectored immunocontraception (pp. 24–30), in R. M. Timm and A. C. Crabb, eds., *Proceedings of the Seventeenth Vertebrate Pest Conference* (Davis: University of California, 1996).

42. K. Williams et al., *Managing Vertebrate Pests: Rabbits* (Canberra: Australian Government Printing Office, 1995).

43. C. K. Williams, Development and use of virus-vectored immunocontraception, *Reproduction, Fertility and Development* 9 (1997):169–178.

44. The disease is also called rabbit calicivirus disease (RCD).

45. Williams et al., *Managing Vertebrate Pests.*

46. Neave, *Rabbit Calicivirus Disease Program.*

47. C. H. Tyndale-Biscoe, Virus-vectored immunocontraception of feral mammals, *Reproduction, Fertility and Development* 6 (1994):281–287.

48. C. K. Williams reviews several of these considerations in Development and use of virus-vectored immunocontraception; C. H. Tyndale-Biscoe discusses some of the social and ethical considerations in Vermin and viruses: Risks and benefits of viral-vectored immunosterilisation, *Search* 26 (1995):239–244.

49. L. Alphey et al., Malaria control with genetically manipulated insect vectors, *Science* 298 (2002):119–121.

50. M. Crichton, *Jurassic Park* (New York: Ballantine Books, 1991).

Chapter 11. Planetary Stewardship

1. F. Galton, The first steps toward the domestication of animals, *Transactions of the Ethnological Society of London, N.S.* 3 (1865):122–138. Cited in J. Clutton-Brock, A Natural History of Domesticated Mammals (Austin: University of Texas Press, 1989), p. 10.

2. *Homo erectus* is generally credited with being the first hominid to use fire, but the evidence is rather difficult to interpret. For a general reference, see "*Homo Erectus*" in the 2003 Encyclopedia Britannica, or go to *http://concise.britannica.com/ebc/article?eu=392583.*

3. H. Walter and S. Breckle, *Ecological Systems of the Geobiosphere* (Berlin: Springer-Verlag, 1985).

4. B. Keen, *The Life of Admiral Christopher Columbus* (New Brunswick, N.J.: Rutgers University Press, 1959).

5. R. A. Anthes, Enhancement of convective precipitation by mesoscale variations in vegetative covering in semiarid regions, *Journal of Climate and Applied Meteorology* 23 (1984):540–553.

6. Noah Webster (1799) from J. Kittredge, *Forest Influences* (New York: McGraw-Hill, 1948).

7. A. C. Becquerel, *Des climats et de l'influence qu'exercent les sols boisés et non boisés,* (Paris: Didot, 1853); see also F. B. Hough, Report upon forestry, *Proceedings of the American Association for the Advancement of Science* (Salem: Salem Press, 1878).

8. J. Shukla and Y. Mintz, Influence of land-surface evapotranspiration on the earth's climate, *Science* 215 (1982):1498–1501; M. D. Schwartz and T. R. Karl, Spring phenology: Nature's experiment to detect the effect of "green-up" on the surface maximum temperatures, *Monthly Weather Review* 118 (1990):883–890; M. M. Fennessy et al., The simulated Indian monsoon: A GCM sensitivity study, *Journal of Climate* 7 (1994):33–41; B. P. Hayden, Ecosystem feedbacks on climate at the landscape scale, *Philosophical Transactions of the Royal Society of London, Series B,* 353 (1998):5–18.

9. Hayden, Ecosystem feedbacks.

10. R. E. Dickinson and A. Henderson-Sellers, Modelling tropical deforestation: A study of GCM land-surface parameterizations, *Quarterly Journal of the Royal Meteorological Society* 114 (1988):439–462.

11. The first experiment employed a relatively coarse spatial resolution GCM. It used a two-layered representation of the hydrology but did not include a plant canopy. A. Henderson-Sellers and V. Gornitz, Possible climatic impacts of land cover transformations, with particular emphasis on tropical deforestation, *Climatic Change* 6 (1984):231–258.

12. J. P. Malingreau and C. J. Tucker, Large-scale deforestation in the southern Amazon Basin of Brazil, *Ambio* 17 (1988):49–55.

13. Y. Xue and J. Shukla, The influence of land surface properties on Sahel climate: Part I, Desertification, *Journal of Climate* 6 (1993):2232–45.

14. G. B. Bonan, D. Pollard, and S. L. Thompson, Effects of boreal forest vegetation on global climate, *Nature* 359 (1992):716–718.

Index

Page numbers in *italics* refer to illustrations.